Tim White
612-196-5911

The publisher gratefully acknowledges the generous contribution to this book provided by the General Endowment Fund of the University of California Press Foundation.

TWILIGHT
OF THE
MAMMOTHS

ORGANISMS AND ENVIRONMENTS

Harry W. Greene, Consulting Editor

TWILIGHT
OF THE
MAMMOTHS

ICE AGE EXTINCTIONS
AND THE REWILDING
OF AMERICA

Paul S. Martin

Foreword by Harry W. Greene

UNIVERSITY OF CALIFORNIA PRESS

BERKELEY   LOS ANGELES   LONDON

University of California Press, one of the most distinguished university presses in the United States, enriches lives around the world by advancing scholarship in the humanities, social sciences, and natural sciences. Its activities are supported by the UC Press Foundation and by philanthropic contributions from individuals and institutions. For more information, visit www.ucpress.edu.

University of California Press
Berkeley and Los Angeles, California

University of California Press, Ltd.
London, England

© 2005 by The Regents of the University of California

Library of Congress Cataloging-in-Publication Data

Martin, Paul S. (Paul Schultz), 1928–.
  Twilight of the mammoths : ice age extinctions and
the rewilding of America / Paul S. Martin.
      p.   cm.— (Organisms and environments ; 8)
  Includes bibliographical references and index.
  ISBN 0-520-23141-4 (acid-free paper)
  1. Extinction (Biology) I. Title. II. Series.
QE721.2.E97M34   2005
560'.1792'097—dc22   2005005745

Manufactured in the United States of America

13   12   11   10   09   08   07   06   05
10   9   8   7   6   5   4   3   2

Dedicated to my wife,
Mary Kay O'Rourke

I listen to a concert in which so many parts are wanting. . . . I seek acquaintance with nature,—to know her moods and manners. Primitive nature is the most interesting to me. I take infinite pains to know all the phenomena of the spring, for instance, thinking that I have here the entire poem, and then, to my chagrin, I hear that it is but an imperfect copy that I possess and have read, that my ancestors have torn out many of the first leaves and grandest passages, and mutilated it in many places.

Henry David Thoreau, *The Journal of Henry David Thoreau*

Contents

Illustrations

TABLES

FIGURES

Foreword

Twilight of the Mammoths: Ice Age Extinctions and the Rewilding of America is the eighth volume in the University of California Press's series on organisms and environments, whose unifying themes are the diversity of plants and animals, the ways in which they interact with each other and with their surroundings, and the broader implications of those relationships for science and society. We seek books that promote unusual, even unexpected, connections among seemingly disparate topics, and we encourage projects that are distinguished by the unique perspectives and talents of their authors. Previous volumes have spanned the ecology of Arizona grasslands, Seri ethnoherpetology, and the biology of Gila monsters.

Twilight of the Mammoths is an insightful, engrossing account of the end of the famous Pleistocene ice ages and of the first colonization by humans from Asia of the New World. Told in the form of a personal journey, Paul Martin's book covers his own boyhood bird-watching, graduate work in evolutionary biology, and a distinguished academic career; it culminates in a daring plan to truly rewild North America. This is a story of science in action, of arduous fieldwork and exciting discoveries, of intellectual puzzles and clashing theories. It is also a work of high-stakes advocacy, in which Martin marshals the evidence for his controversial theory that humans, within a remarkably short time after our arrival, caused the extinction of more than 30 genera and 40 species of large mammals. More than that, Martin aims to convince us to take the long view, to learn from our collective past so that we can enrich the future of life on earth. He wants us to regard feral horses and burros as repatriated natives rather than introduced pests, to welcome Asian elephants as surrogates of extinct proboscideans. He challenges us with these and other bold proposals to ask: Why did we in North America inherit such an impoverished mammal fauna? What kind of world will our children inherit?

Imagine a Serengeti-like vista on the North American Great Plains, a region that actually housed and fed not only roughly 30 million bison, but also countless individuals of dozens of other large mammal species. Realize that we are not conjuring an ancient fantasy here, that *Twilight of the Mammoths* is not about some furry version of *Jurassic Park*, only vaguely based on reality. Until about 13,000 years ago—just 130 centuries, or a few times as long ago as the pharaohs reigned over Egypt—western North America really did harbor a megafauna that surpassed Africa's modern biota in species richness. There were several species of wild horses and camels, an armadillo relative the size of a small car, and giant ground sloths, as well as short-faced bears, dire wolves, saber- and scimitar-toothed cats. There were somewhat larger versions of our contemporary cheetahs and lions, and there were several species of mastodons and mammoths, relatives of living elephants. All of those large New World mammals are gone now, vanished in a heartbeat of geological time, and yet almost all of the plants and smaller species that lived with them persist today. Some of those surviving organisms, like the Osage orange tree, are still here only because we have taken on the ecological roles of extinct megafauna, in this case seed dispersal; one of our surviving ungulates, the pronghorn, is capable of locomotor feats that only make sense in the presence of a high-speed predator like the now-extinct North American cheetah. Today the sole substantial remnant of what was once a global Pleistocene megafauna is in Africa, a rapidly changing landmass on which people are killing each other in tragic numbers in response to shrinking resources. Throughout the world, large vertebrates have been reduced to life in fragmentary habitats and often to dangerously low population levels if they are to survive and make long-term evolutionary adjustments to environmental change.

Paul Martin is a true visionary, a time traveler who thinks across the expanses of prehistoric millennia and entire continents with the ease with which most of us locate a car we parked yesterday. His professional accomplishments include a classic work on amphibian and reptile biogeography, pioneering studies of palynology and plant ecology, and an illustrious legacy of former students and research associates. Paul is a gifted storyteller, a self-described lover of "tempting diversions," and I doubt anyone else could make standing chest-deep in extinct sloth dung sound so magical. Soon after identifying the pollen of an abundant local plant in that 13,000-year-old manure, this voracious naturalist tries out the leaves and flowers of globe mallow on his own digestive tract. Paul is by temperament affable rather than cantankerous, and here he generously confronts the full panoply of his critics. He faces squarely "overchill" and "overill," the alternative hypotheses that climate or disease killed off the Pleistocene megafauna, and he candidly confronts charges that the overkill theory reflects cultural insensitivity.

Twilight of the Mammoths is an intellectual detective story, one that remains in part controversial but that also resonates marvel and hope. When I visited Nairobi National Park, surrounded for the first time in my life by thousands of large wild herbivores and alert to the unseen presence of their predators, I experienced a surprising nostalgia for our own, largely extinct North American savanna faunas. Now Paul Martin's life's work challenges all of us who care about nature to be hopeful rather than sad, to think very big. In that spirit he gives us the thrill of biological exploration, the facts as they have emerged thus far, and some of the problems remaining to be solved. His legacy amounts to a profound challenge—that we face up to the near cessation of large vertebrate evolution because of habitat destruction and human persecution. Martin asks us to accept that humans are, willy nilly, in charge of the fate of wilderness and to consider an optimistic alternative to extinction. In this captivating book he says, look at the ecological processes and evolutionary potential that we've so recently lost—in so far as we can, let's bring them back!

Harry W. Greene
February 2005

Acknowledgments

For four decades I have been traveling in my imagination to and from the near-time wilderness to the prehistoric entire Earth still inhabited by ground sloths, mammoths, and other extinct beasts. Some call it the late Quaternary, the youngest of a series of glacial cycles that began 1.8 (others would say 2.3) million years ago and has yet to end.

Many fine fossil localities suitable for sampling near time or at least capturing its flavor lie within a day's drive of Tucson. Besides ranchers, land managers, and backpackers with a keen eye for the environment, my best scouts for important habitats or new fossil localities, including fossil packrat middens, have been students and colleagues conducting research of their own. They have guided me to informative fossil deposits and generously shared their insights, often at lunch on the patio of the Desert Laboratory of the University of Arizona or at fiery "potluck seminars," informal gatherings once held by dedicated students and footloose faculty from various corners of the university known as the Menudo Society. The seminars originated at the hand of former University of Arizona Professor Alan Solomon, his students, and his mentor, the late ecologist Murray Buell of Rutgers University.

For field, lab, editorial, and interpretive insights I thank a number of former students, colleagues, and visitors to the Desert Laboratory. In addition to others mentioned in the text, I thank David P. Adam; Larry Agenbroad; Wanda Agenbroad; Martha Ames; Robyn Andersen; Connie Barlow; Larry Belli; Cynthia Bennett; Peter Bennett; Julio Betancourt; George

Billingsly; Russ Boulding; Jan Bowers; Diane Boyer; Georgie Boyer; the late Jim Boyer; Bob Brumbaugh; Tony Burgess; Dave and Lida Pigott Burney; Steve Carothers; Ken Cole; Nick Czaplewski; Owen K. Davis; Russell Davis; Jared Diamond; Bill Dickinson; Steve Emslie; the late Robert Euler; George Ferguson; Stuart Fiedel; Claire Flemming; Karl Flessa; George C. Frison; Richard Gillespie; Alan Gottesfeld; Russ Graham; Harry W. Greene; Dale Guthrie; Daniel A. Guthrie; C. Vance Haynes; Gary Haynes; William B. Heed; Donna Howell; Jeff Ingram; Bonnie Fine Jacobs; Lewis Jacobs; Helen James; Roy Johnson; Charles E. Kay; Gerald Kelso; Lloyd F. Kiff; the late Fran Bartos King; Jim King; Stan Kryzanowski; Steve Kuhn; Donna LaRocca; Cynthia Lindquist; Everett Lindsay; Ernest Lundelius; Ross MacPhee; Vera Markgraf; Andrea Martin; Andrew G. Martin; Marianne W. Martin; Neil M. Martin; Thomas C. Martin; Edgar J. McCullough; H. G. McDonald; Emily Mead; Jim I. Mead; Peter J. Mehringer; Eric Mellink; Kathy (Kik) Moore; Mary Ellen Morbeck; James E. Mosimann; Gary P. Nabhan; Phil R. Ogden; John Olsen; Storrs Olson; Mary Kay O'Rourke; the late Wes Peirce; Arthur M. Phillips III; Barbara Phillips; Greg Pregill; Vernon Proctor; Ron Pulliam; Brian Robbins; Eleanora (Norrie) Robbins; Guy Robinson; the late Ike Russell; Jean Russell; Jeff Saunders; Louis Scott; Pat Shipman; Jennifer Shopland; David W. Steadman; Mary Stiner; Todd Surovell; Chris Szuter; Jean Turner; Ray Turner; Sandra Turner; Thomas Van Devender; Nicole Waguespack; Alan Walker; Jim Walters; Peter Warren; Robert Webb; and David Western.

I am deeply grateful to Doris Kretschmer, executive editor at the University of California Press, and to Kate Warne and my peerless developmental editor, Lynn Stewart Golbetz. Without their efforts, this book might never have seen the light of day.

Attitudes toward, interest in, and fascination with nature are shaped long before adulthood. I was very fortunate to be born to farm-raised, nature-loving, college-educated parents. As a veterinarian my Dad specialized in treating dairy cattle for bovine mastitis. Various uncles and aunts farmed, and one operated a small slaughterhouse. As a teenager during World War II, I worked summers for my uncle Wayne, the last of seven generations of the Schultz family to own "Scholtop" farm in upper Montgomery County, Pennsylvania.

Eastern Pennsylvania provided plenty of cover for small game. We lived at the edge of the borough of West Chester. After school I hiked across lots, fields, and woods in search of water birds on the West Chester Reservoir and along Brandywine Creek, habitats that would have attracted proboscideans, had there been any. During hunting season with relatives I shot pheasants and rabbits, and in winter we trapped a few muskrats. With members of the West Chester Bird Club I visited Hawk Mountain, Pennsylvania; Cape May, New Jersey; and Bulls Island, outside Charleston, South Carolina, habitats magnetically attractive in season to birds and thus to bird-watchers.

Last and not least are unforgettable acts of kindness that after five decades remain fresh in my mind. Bill Dilger and Bob Dickerman loaded me, paralyzed, on a flight from Ithaca to Cornell's Medical Center in Manhattan. Experienced with and sensitive to such matters from his own traumatic life, David Kirk brought my distraught partner in those days into the comfort of his family circle. She and her parents and mine, and my cousin Alma, kept my spirits up through a neurological impairment represented by specialists as potentially terminal. After remission two months later, the outcome was a chronic if minor handicap, one that opened a door for explorations of near time.

PROLOGUE

Imagine a world with only half the variety of large animals that we know today.

Imagine an Africa with hyenas but no lions, an Australia with wombats but no koalas, a North America with elk but no bison.

Imagine zoos and televised nature programs featuring rhinos without hippos, giraffes without gorillas, zebras without camels, leopards without cheetahs. The missing animals simply do not exist; we know them only from fossils.

Without realizing it, we are in exactly this situation today. In what paleontologists have begun to call "near time," the last 50,000 years, datable by radiocarbon, the world lost half of its 200 genera of large mammals (those weighing more than 45 kilograms or 100 pounds). Beyond the living bears, bison, deer, moose, and other large mammals familiar to us now, an additional 30 genera and over 40 species lived in North America, and even more in South America. Most of the Western Hemisphere's charismatic large mammals no longer exist. As a result, without knowing it, Americans live in a land of ghosts.

Some of these great creatures—the extinct megafauna—appear in popular museum displays in our large cities. Even the names of others are utterly unfamiliar to most of us. North America lost mastodons, gomphotheres, and four species of mammoths; ground sloths, a glyptodont, and giant armadillos; giant beavers and giant peccaries; stag moose and dwarf antelopes; brush oxen and woodland musk oxen; native camels and horses; short-faced bears, dire wolves, saber-toothed and dirk-toothed cats, and an American subspecies of the king of beasts, the lion. After the extinctions, the mean body mass of North American mammals was the lowest it had been in 30 million years (Alroy 1999).

The survivors of the big wipeout are those large animals familiar to us now, such as bison, brown (grizzly) bears, cougars (mountain lions), deer, elk (wapiti), moose, musk oxen, and pronghorns. Most people regard these as defining "wild America." They do not. To give so little attention to the dozens of big animals we have lost so recently simply sells North America short. Before extinction of our native big mammals, the New World had much more in common with an African game park than most of us realize.

South America also lost heavily. Extinction struck many species of ground sloths, one monster weighing over 4,500 kilograms (10,000 pounds) (Fariña, Vizcaíno, and Bargo 1997). The biggest native herbivore in the New World tropics today is Baird's tapir, which may reach 225–300 kilograms (about 500–650 pounds).

Australia lost giant animals of its own. Though not as massive as the largest in the Americas, they included giant wombat-like creatures the size of rhinos, giant kangaroos larger than any of the living kangaroos, many other large marsupials, and even some oversized koalas and echidnas. (Echidnas, or spiny anteaters, differ from all other mammals in that their shell-covered eggs are incubated and hatched outside the mother's body.)

If we could travel back just those 50,000 years—a third of the age of our species in Africa, but a mere 1/80,000th of the roughly 4.5-billion-year age of the Earth—we would find ourselves in a "Quaternary zoo" far more spectacular and much richer in species of large mammals than any zoo that exists today.

Any fan of modern wombats—muscular 4-foot-long marsupials resembling badgers with elongated koala faces—would delight in seeing diprotodons, which looked like one-ton wombats. The diprotodons and their entire family suffered extinction over 40,000 years ago.

Those who enjoy America's modern armadillos would take particular pleasure in the glyptodonts, another extinct family. The size and overall shape of a giant tortoise or of a Volkswagen "beetle" (Hulbert 2001), glyptodonts were completely armored in bone and had long, muscular tails ending in a club or a mace-like cudgel, presumably used to beat off attackers.

We would gaze in awe at mastodons, mammoths, and gomphotheres, all relatives of modern elephants. The American display would also feature several species of ground sloth, some of them as large as the mammoths. The Shasta ground sloth was about the size of a large black bear. The public would flock to the viewing platform at feeding time, when buckets of this animal's favorite vegetable, mallows in the hollyhock family, would attract patient mothers carrying their young on their backs. It would be harder to feed the biggest ground sloths, elephant-sized *Megatherium,* able to reach high into trees, pulling down branches with their long arms and heavy claws. Some paleontologists suspect they may also have been scavengers, eating carcasses of large dead animals.

The carnivore displays would be equally impressive. The short-faced bear, *Arctodus,* exceeded all living bears in size and probably in speed. The famous saber-toothed cat *(Smilodon)* was about the size of today's African lion, with curved 7-inch-long upper canines, while the canines of the scimitar cat *(Homotherium)* were "only" 4 inches long. Among the other carnivores were a subspecies of lion, *Panthera leo atrox,* as well as the dire wolf, *Canis dirus,* along with the dhole, *Cuon,* a wild dog that survives in Asia.

These are some of the more spectacular mammals we would see in a Quaternary zoo, side by side with our familiar bears, bison, and hippos. If the zoo included birds and reptiles, we would also see New Zealand's moa—10 extinct species of flightless, hairy-looking birds, the largest of them bigger than ostriches. Among the wondrous Australian giants were fearsome monitor lizards weighing up to 150 kilograms (330 pounds); two terrestrial crocodiles; and a giant extinct python.

All of these animals were present on the planet until well into the lifetime of our own species. Why are they gone from the Earth today? In this book I argue that virtually all extinctions of wild animals in the last 50,000 years are anthropogenic, that is, caused by humans. To get our history right we need to know more about the extinctions of near time. And we need to give thought to reversing prehistoric extinctions when we have the chance. That leads us to the most controversial vista of them all, the contemplation of bringing back the elephants and representatives of other lineages that evolved over tens of millions of years in the Americas.

ONE

DISCOVERING THE LAST LOST WORLD

> To produce a mighty book you must choose a mighty
> theme . . . to include the whole circle of the sciences, and
> all the generations of whales and men and mastodons.
> **Herman Melville, *Moby-Dick***

Career paths are notoriously unpredictable, and I never imagined that
mine would lead me to focus on prehistoric megafaunal extinction. Beginning in 1948 at Rancho del Cielo, a cloud forest just within the Tropic
of Cancer in eastern Mexico, I collected birds for ornithologist George M.
Sutton and other vertebrates for the University of Michigan's Museum
of Zoology. The best part was the forest of tall, dense trees rising to 100
feet: sweet gum, many species of oak unknown north of the border, *Mag-
nolia, Podocarpus*, redbud, a few scattered palms, and a wealth of tank
bromeliads, a home for frogs and lungless salamanders. Had I not suffered a handicap from a bout with polio in 1950, I might never have
turned my attention to peat, rich in fossil pollen, being studied by
botanists in the postglacial lakes around Ann Arbor. Then, in 1955, I
learned from ecologist Ed Deevey at Yale University how to extract, identify, and count fossil pollen. From these counts one could learn what happened to plants after the glaciers melted away.

For Ed, the biogeography of the Pleistocene (the last ice age, 1.8 million years ago to 10,000 years ago), with all its glacial and interglacial
changes in climate, was the key to understanding modern plant and animal distributions. Ed took cores of organic sediment from lake beds
and counted samples of the fossil pollen, spores, and copepods (minute
aquatic crustaceans) they contained. He could date these remains by

Willard F. Libby's then-new radiocarbon method; scientists could now refer to "Libby time" (roughly the last 40,000 years, the period for which radiocarbon dating is most effective). Magically, the fossil pollen record in sediments cored from New England lakes told of the comings and goings of treeless tundra and of spruce, fir, jack pine, and other trees as the climate warmed, the glaciers melted and on occasion readvanced, and eventually the ice-margin boreal vegetation yielded to today's deciduous forest. It even gave clues to the fate of animal species during this period. For example, fossil pollen counts plotted in percentages as a diagram associated with bones of mastodons indicated that they vanished around the time that, according to the fossil pollen counts, spruce gave way to pine. Some paleontologists thought that the change in tree cover from boreal conifers to temperate hardwoods might help explain mastodon extinction. Then they dropped the extinction question and returned to their primary interests, vertebrate anatomy, evolution, and geochronology. Nobody bothered to study the extinctions. Somehow the Pleistocene megafauna, big as it was, remained out of sight and out of mind.

In the winter of 1956, my wife and I found ourselves raising our children in a tenant farmhouse we rented from Anatole Cecyre, a French Canadian dairy farmer outside Chateauguay, Quebec. I commuted to a postdoctoral fellowship at the Université de Montréal, working on a pollen record of late-glacial climatic change and offering a seminar on Quaternary biology held jointly with McGill University.

Identifying and counting pollen grains can become monotonous. In a break from the microscope one day, I skimmed through George Gaylord Simpson's monumental *Classification of Mammals,* a long list of genera organized taxonomically. Malcolm McKenna and Susan Bell of the American Museum of Natural History have recently revised this classic tome. They recognize 5,158 mammalian genera, of which 4,075, about 80 percent, are extinct (McKenna and Bell 1997). The large number of mammal extinctions is to be expected, because the list embraces the end of the Mesozoic (the era that began about 250 million years ago) and the entire Cenozoic (the era that began about 65 million years ago and continues to today).

As a diversion that snowy subzero weekend, a diversion that fit right into the seminar on Quaternary biology, I began to plot all the late-Quaternary megafaunal extinctions listed by Simpson against those that had taken place earlier in the Cenozoic. After two days I was stunned by what I found. In the Miocene (starting 24 million years ago) and the Pliocene (starting five million years ago), many mammals of all sizes turned over (i.e., evolved and went extinct). But at the end of the Quaternary, the pattern of extinctions in North America became very strange. It was the large terrestrial mammals that disappeared, and those for which radiocarbon dates were available did so suddenly.* Large marine mammals, on the other hand—the whales, dolphins, and pinnipeds—had been hard hit by extinctions in the Miocene and Pliocene but survived the Quaternary virtually intact. So did most of the continental small mammals (the shrews, moles, rats, and mice).

The major event in the Quaternary was the extinction of the large terrestrial mammals. Extinction also doomed their endemic species of parasites and commensals. (Commensals are species that benefit from accompanying other species without necessarily harming them.) For example, near-time extinction of internal parasites of ground sloths (Schmidt, Duszynski, and Martin 1992) or reduction in the number of species of cowbirds, magpie-type corvids, and dung beetles can be accepted as secondary, given the apparent dependence of these species on large mammals (Steadman and Martin 1984). Along with mammoths and ground sloths, the late-Quaternary extinctions also involved avian scavengers (such as condors), commensals (such as the Thick-knees, or Stone Curlews, of the Old World and tropical America), or guardians of the big mammals that eat their external parasites, such as tickbirds, which fly away if alarmed, alarming their host (Steadman and Martin 1984).

In the last 10,000 years, after most of the extinctions of the big mammals on continents, many small mammals, birds, reptiles, and land snails vanished. Small animal extinction and dwarfing, as well as extinction of some large mammals and birds, happened on oceanic islands, such as those in the West Indies and New Zealand, and on those in the Mediterranean. Thousands of small islands, especially in the remote Pacific, saw extinctions of small animals, especially birds and endemic species of land snails. The deep water surrounding these islands precluded any connection to each other or to the mainland, even when the sea level dropped

---

*John Alroy (1999, 2001) has looked at the fossil record in much greater detail. His analysis shows that late-Pleistocene extinctions in North America are quantitatively unlike any of the changes seen earlier in the Cenozoic. Only in the late Pleistocene is heavy extinction focused so strongly on large mammals.

Throughout this book, I call vertebrates large if they weigh over 45 kilograms (100 pounds), the size of a small adult human or an adult pronghorn. Some scholars define large mammals as those weighing over 1 kilogram (Alroy 1999, 2001). Just where the boundary is located does not alter the overall pattern: in near time on the continents, far more large mammals went extinct than small ones.

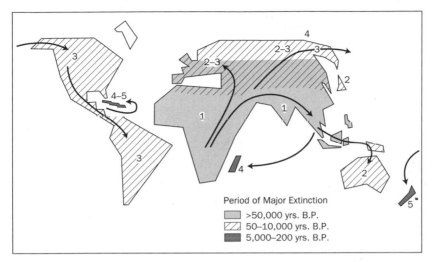

Figure 1. Map showing sequence of extinctions. Arrows indicate direction of human dispersals; numbers indicate order of human settlement. Adapted from Martin 1970, © American Institute of Biological Sciences.

by 400 feet or more, as it last did during the height of continental glaciation around 18,000 years ago.

In contrast, islands on the continental shelf, including Britain, Sri Lanka, Java, Sumatra, and Trinidad, were connected to continents when the sea level dropped.* Their faunas were much less vulnerable to prehistoric loss. Often much smaller than their continental relatives, the animals that evolved on oceanic islands included the dwarf mammoths of Santa Rosa and San Miguel islands off the California coast; in the Mediterranean, the dwarf elephants of Crete and the dwarf hippo of Cyprus; in the West Indies, the dwarf ground sloths of Cuba and Hispaniola (Haiti and the Dominican Republic). The tropical islands of the remote Pacific, some of them quite small, hosted 2,000 taxa of flightless rails. Apparently depending on when humans (often accompanied by Pacific rats) first arrived, many of the island endemics suffered prehistoric extinction. Pacific

*Islands surrounded by shallow water of the continental shelves, less than 120 meters (400 feet) deep, would not emerge long enough to evolve highly endemic species. These shelf islands disappear as rising interglacial sea levels shrink and eventually drown them. Islands artificially formed by impoundment of rivers experience comparable extinctions of larger animals and artificial increases in smaller ones. In contrast, deep-water islands could be colonized only by species surviving water transport or by ancient detachment from a continent such as Gondwanaland. They support faunas that are impoverished but rich in endemics, such as Jamaica's extinct giant rodents, Madagascar's extinct giant lemurs, and Sulawesi's extinct dwarf elephants.

Phanerozoic Eon (543 mya to the present)	Cenozoic Era (65 mya to the present)	Quaternary (1.8 mya to the present)	Holocene (10,000 years ago to the present)	North American Land Mammal Ages
			Pleistocene (1.8 mya to 10,000 years ago)	**Near Time (the last 50,000 years)** Rancholabrean (250,000 years ago to the present) Irvingtonian (1.8 mya to 250,000 years)
		Tertiary (65 to 1.8 mya)	Pliocene (5.3 to 1.8 mya)	Blancan (4.7to 1.8 mya)
			Miocene (23.8 to 5.3 mya) Oligocene (33.7 to 23.8 mya) Eocene (54.8 to 33.7 mya) Paleocene (65 to 54.8 mya)	
	Mesozoic Era (248 to 65 mya)	Cretaceous (144 to 65 mya)† Jurassic (206 to 144 mya) Triassic (248 to 206 mya)		
	Paleozoic Era (543 to 248 mya)	Permian (290 to 248 mya)†		
		Carboniferous (354 to 290 mya)	Pennsylvanian (323 to 290 mya) Mississippian (354 to 323 mya)	
		Devonian (417 to 354 mya)† Silurian (443 to 417 mya) Ordovician (505 to 443 mya)† Cambrian (570 to 505 mya)		

Figure 2. The geologic time scale. Note that Near Time comprises a tiny and recent portion of the Earth's history, underscoring how close to our own time was the world populated by an array of large mammals. "mya" = million years ago; NA = North America; † = mass extinction. Adapted from McKenna and Bell 1997 (© Columbia University Press); Museum of Paleontology, University of California, Berkeley; *Merriam-Webster's Collegiate Dictionary,* 11th ed.

rats *(Rattus exulens)* may have been the agents of over a hundred extinctions of ground-nesting birds in New Zealand and of endemic land snails on thousands of Pacific islands.

Significantly, no avian extinctions have been detected in the Galapagos, where fossil cave faunas sampled by David Steadman (Steadman and Zousmer 1988) on various islands are made up entirely of living species, including species of Darwin's Finches. Extinction of the endemic Galapagos rat, *Megaoryzomys,* coincides with historic human contact. On a visit to the island of Santa Fe, Dave and his field assistants searched for Pleistocene sediments and fossils without success. They found a rich fossil record in caves on Floreana, but it was apparently not old enough to include extinct species.

While not the first to be intrigued by these patterns, I was among the first to compare near-time extinctions between continents and continental extinctions with those on oceanic islands. Radiocarbon dating made such comparisons possible. Moreover, vertebrate paleontologists have greatly improved and refined the fossil records of mammals, large and small. After a false start early in the twentieth century by paleontologists such as O. P. Hay, who thought that numerous extinctions of mammals took place early in the Quaternary, paleontologists and geologists came to recognize that megafaunal extinctions in America happened mainly around the end of the last glacial episode, at the end of the Quaternary. This was about 13,000 years ago or less. The extinctions were not only of species and genera, but also of higher taxonomic categories, such as families and occasionally an order. The evolution of a family of mammals normally takes tens of millions of years. Some paleontologists argued fatalistically that the late-Quaternary extinctions were inevitable; their time had come. But so many, so suddenly, in so many corners of the world, and involving established lineages of large mammals on continents? Something strange had happened. What was it?

"Blighted" is the right word for the animal kingdoms of America and Australia after the near-time extinctions ran their course. Globally, extinction was the fate of about half of the genera of large terrestrial mammals known to have existed on the continents at the time. They were soon followed by thousands of species or taxa (taxonomic categories, in this case distinct populations) of island birds and land snails. Prehistoric extinctions swept the remote corners of the Pacific, including Hawaii, the Marquesas, Rapanui (Easter Island), and New Zealand.

In near time, North America lost more genera than it had in the preceding 1.8 million years. Table 1 lists the living and extinct large mammals

TABLE 1 *Late Quaternary Extinct and Living Species of Large (>45 Kilograms) Land Mammals, Western North America and Northern Mexico*

Classification	Common name
Xenarthra	
†*Glyptotherium floridanum*	glyptodont
†*Paramylodon harlani*	big-tongued ground sloth
†*Megalonyx jeffersonii*	Jefferson's ground sloth
†*Nothrotheriops shastensis*	Shasta ground sloth
Carnivora	
*Canis *dirus*	dire wolf
Canis lupus	gray wolf
Ursus americanus	black bear
Ursus arctos	brown (grizzly) bear
†*Arctodus simus*	giant short-faced bear
†*Smilodon fatalis*	saber-toothed cat
*Panthera leo *atrox*	American lion
Panthera onca	jaguar
†*Miracinonyx trumani*	American cheetah
Puma concolor	mountain lion
Proboscidea	
†*Mammut americanum*	American mastodon
†*Mammuthus columbi*	Columbian mammoth
†*Mammuthus exilis*	dwarf mammoth
†*Mammuthus primigenius*	woolly mammoth
Perissodactyla	
*Equus *conversidens*	Mexican horse
*Equus *occidentalis*	western horse
*Equus *spp.*	extinct horses or asses
*Tapirus *californicus*	tapir
Artiodactyla	
†*Camelops hesternus*	western camel
†*Hemiauchenia macrocephala*	long-legged llama
**Mylohyus nasutus*	long-nosed peccary
**Platygonus compressus*	flat-headed peccary
Odocoileus hemionus	mule deer
Odocoileus virginianus	white-tailed deer
**Navahoceros fricki*	mountain deer
Rangifer tarandus	woodland caribou
Alces alces	moose, moose deer
Cervus elaphus	wapiti, elk

TABLE 1 *continued*

Classification	Common name
Antilocapra americana	pronghorn
*Oreamnos *harringtoni*	Harrington's mountain goat
Oreamnos americanus	mountain goat
Ovis canadensis	bighorn
Euceratherium collinum	shrub ox
Bootherium bombifrons	bonnet-headed musk ox
Bison bison	bison
*Bison *spp.*	extinct bison

SOURCE: After Martin and Szuter 1999. Courtesy Blackwell Publishing.
†Extinct genus.
*Extinct species. The more common taxa have terminal radiocarbon dates around 13,000 calendar years ago (Stuart 1991).

known in western North America and northern Mexico in near time. Table 2 lists the large mammals of North America north of Mexico over the last two million years and shows the concentration of extinctions in the late Quaternary, with its distinctive Rancholabrean fauna. Table 3 lists all living and extinct large land mammals of near time found throughout the world.

It is well worth examining in more detail the kinds of animals that went extinct in the late Quaternary and the geographic regions that were affected. The following sketch treats some of the more common large animals eliminated by extinctions in the last 50,000 years. (For more details and illustrations see E. Anderson 1984; Hulbert 2001; Kurtén 1988; Kurtén and Anderson 1980; Lange 2002; MacPhee 1999; Martin and Klein 1984; Murray 1991; Steadman n.d.; and Sutcliffe 1985.)

AFRICA AND EURASIA: THE CONTROLS

The large-mammal faunas of Africa and Eurasia escaped severe extinctions over the last two to five million years. Africa suffered no obvious pulse or burst of extinctions of megafauna to match those in near time on other landmasses; in the end it lost less than 10 percent of its megafauna. The losses in northern Eurasia were not aggregated or spontaneous, as in America and Australia, but more measured, extending over 70,000 years (Stuart 1999). They included straight-tusked elephants and woolly mammoths, naked and woolly rhinos, hippos, giant deer, and cave bears *(Ursus spelaeus).* Cave lions and spotted hyenas, close relatives of the

TABLE 2 *Large (>45 Kilograms) Pliocene, Pleistocene, and Holocene Mammals of North America North of Mexico*

	Blancan				Irvingtonian			Rancholabrean		Holocene
	1, 2	3	4	5	Early	Middle	Late	Early	Late	
Stage duration (millions of years)	1.0	0.5	0.5	0.2	0.9	0.4	0.2	0.2	0.1	0.01
XENARTHRA										
Dasypus *sp., giant armadillo			•	•	•	•	•	•	•	
†*Holmesina*, northern pampathere			•	•	•	•	•	•	•	
†*Pachyarmatherium*, glyptodont							•			
†*Glyptotherium*, glyptodont			•	•	•	•	•	•	•	
†*Eremotherium*, giant ground sloth					•	•	•	•	•	
†*Nothrotheriops*, Shasta ground sloth				•	•	•	•	•	•	
†*Megalonyx* spp., Jefferson's ground sloth	•	•	•	•	•	•	•	•	•	
†*Paramylodon*, big-tongued ground sloth		•	•	•	•	•	•	•	•	
CARNIVORA										
†*Borophagus* sp., plundering dog	•	•	•	•	•					
Canis *spp., dire wolf, others				•	•	•	•	•	•	•

TABLE 2 continued

†*Protocyon*, Troxell's dog		•				
Ursus spp., bears	•	•	•	•	•	•
*Tremarctos, Florida cave bear	•	•	•	•	•	
†*Arctodus* spp., short-faced bears		•	•	•	•	
†*Chasmaporthetes*, hunting hyena	•	•				
†*Megantereon*, western dirk-toothed cat	•	•				
†*Smilodon*, saber-toothed cat		•	•	•	•	
†*Ischyrosmilus*, Idaho saber-toothed cat	•					
†*Homotherium*, scimitar cat		•	•	•	•	
†*Dinofelis*, false saber-toothed cat	•	•				
*Panthera spp., American lion, jaguar		•	•	•	•	•
†*Miracinonyx*, American cheetah	•	•	•	•	•	
Puma, cougar, panther		•	•	•	•	•
RODENTIA						
†*Procastoroides*, large beaver	•					
†*Castoroides*, giant beaver		•	•	•	•	

TABLE 2 *continued*

	Blancan				Irvingtonian			Rancholabrean		Holocene
	1, 2	3	4	5	Early	Middle	Late	Early	Late	
Stage duration (millions of years)	1.0	0.5	0.5	0.2	0.9	0.4	0.2	0.2	0.1	0.01
†*Neochoerus*, Pinckney's capybara					•	•	•	•	•	
*Hydrochoerus sp., Holmes's capybara							•	•	•	
PROBOSCIDEA										
†*Mammut*, American mastodon	•	•	•	•	•	•	•	•	•	
†*Stegomastodon*, stegomastodon		•	•	•	•					
†*Rhyncotherium*, rhynchothere	•	•	•	•						
†*Cuvieronius*, gomphothere			•	•	•	•	•	•	•	
†*Mammuthus* spp., mammoths				•	•	•	•	•	•	
SIRENIA										
†*Hydrodamalis*, Steller's sea cow	•	•	•	•	•	•	•	•	•	•
Trichechus, manatee	•	•	•	•	•	•	•	•	•	•

TABLE 2 *continued*

PERISSODACTYLA						
†*Cormohipparion*, extinct equid	•	•				
†*Nannipus*, gazelle-horse	•	•	•			
†*Plesippus*, extinct equid	•	•	•			
*Equus spp., horses	•	•	•	•	•	•
*Tapirus spp., tapirs	•	•		•	•	•
ARTIODACTYLA						
†*Megalotylopus*, large camelid	•	•				
†*Blancocamelus*, giraffe-camel		•				
†*Titanotylopus*, giant camelid	•	•	•	•		
†*Camelops* spp., camels	•	•	•	•	•	•
†*Hemiauchenia*, llama	•	•	•	•	•	•
†*Palaeolama*, stout-legged llama				•	•	•
†*Mylohyus*, long-nosed peccary	•	•	•	•	•	•
†*Platygonus*, flat-headed peccary	•	•	•	•	•	•
†*Bretzia*, false elk	•					
Odocoileus spp., deer	•	•	•	•	•	•
†*Torontoceras*, extinct large cervid	•					

TABLE 2 continued

Stage duration (millions of years)	Blancan				Irvingtonian			Rancholabrean		Holocene
	1, 2	3	4	5	Early	Middle	Late	Early	Late	
	1.0	0.5	0.5	0.2	0.9	0.4	0.2	0.2	0.1	0.01
†*Navahoceros*, mountain deer									•	
Rangifer, caribou, reindeer				•	•	•	•	•	•	•
Alces spp., moose, broad-fronted moose							•	•	•	•
†*Cervalces*, stag moose								•	•	
Cervus, wapiti (elk)				•	•	•	•	•	•	•
†*Tetrameryx*, four-horned pronghorns					•	•	•	•	•	
†*Hayoceros*, Hay's pronghorn							•			
†*Stockoceros*, four-horned pronghorns							•	•		
Antilocapra, pronghorn								•	•	•
†*Saiga*, saiga (survives in Eurasia)								•	•	
Oreamnos spp., mountain goats								•	•	•
Ovis, bighorn or mountain sheep								•	•	•
†*Euceratherium*, shrub-ox					•	•	•	•	•	
†*Soergelia*, Soergel's ox					•	•	•	•	•	
†*Bootherium*, bonnet-headed musk ox							•	•	•	

TABLE 2 *continued*

†*Praeovibos*, Staudinger's musk ox						•				
Ovibos, musk ox									•	•
*Bison spp., bison species								•	•	•
†*Platycerabos*, flat-horned ox								•		
PRIMATES										
Homo, modern humans									•	•
ORIGINATIONS OF GENERA	—	6	5	6	12	3	7	2	10	0
EXTINCTIONS OF GENERA	0	2	2	6	5	0	4	1	34	1
Total genera	18	24	27	31	37	35	42	40	49	15

SOURCE: After Martin and Steadman 1999, excluding *Homo*. Courtesy Springer Science and Business Media.
†Extinct genus.
*Extinct species.

TABLE 3 *Near-Time Extinct and Living Genera*
of Large Land Mammals, Worldwide (>40 kilograms)
(including genera living on one continent and suffering extinction on another)

Higher category	Extinct Genus	Living Genus
Cohort Marsupialia		
†Family		
Palorchestidae	†*Palorchestes* Au	
†Family		
Thylacoleonidae	†*Thylacoleo* Au	
†Family		
Diprotodontidae	†*Zygomaturus* Au	
	†*Hulitherium* Au (NG)	
	†*Maokopia* Au (NG)	
	†*Diprotodon* Au	
Family Vombatidae	†*Phascolonus* Au	
	†*Ramsayia* Au	
Family Macropodidae		
	†*Propleopus* Au	*Macropus* Au, gray
	†*Protemnodon* Au	kangaroo
	†*Troposodon* Au	*Megaleia* Au,
	†*Congruus* Au	red kangaroo
	†*Sthenurus* Au	
	†*Procoptodon* Au	
Cohort Placentalia		
Order Bibymalagasia	†*Plesiorycteropus* Ma, 5–10 kg	
Order Cingulata		
Family Dasypodidae	†*Propraopus* SA	*Priodontes* SA,
	†*Eutatus* SA	giant armadillo
†Family Pampatheriidae	†*Pampatherium* NA SA	
†Family Glyptodontidae	†*Neothoracophorus* SA	
	†*Hoplophorus* SA	
	†*Lomaphorus* SA	
	†*Panochthus* SA	
	†*Parapanochthus* SA	
	†*Doedicurus* SA	
	†*Plaxhaplous* SA	
	†*Glyptodon* SA	
	†*Heteroglyptodon* SA	

TABLE 3 *continued*

Higher category	Extinct Genus	Living Genus
	†*Glyptotherium* NA	
	†*Neuryurus* SA	
	†*Doedicurus* NA	
Order Pilosa		
Suborder Phyllophaga		
†Family Scelidotheriidae	†*Scelidotherium* SA	
†Family Mylodontidae	†*Mylodon* SA	
	†*Glossotherium* SA	
	†*Paramylodon* NA	
	†*Mylodonopsis* SA	
	†*Lestodon* SA	
	†*Lestodontidion* SA	
†Family Megatheriidae	†*Megatherium* SA	
	†*Eremotherium* NA SA	
	†*Perezfontanatherium* SA	
	†*Nothropus* SA	
	†*Nothrotherium* SA	
	†*Nothrotheriops* NA	
†Family Megalonychidae	†*Valgipes* SA	
	†*Megalonyx* NA SA	
	†*Megalocnus* NA	
	(Cuba), 200 kg	
Order Rodentia		
Family Castoridae	†*Castoroides* NA	
	(giant beaver)	
Family Hydrochoeridae	†*Neochoerus* NA SA	*Hydrochoerus* NA SA, capybara
Order Carnivora		
Family Felidae	†*Homotherium* NA As	*Felis* (Puma) NA SA, mountain lion, cougar
	†*Smilodon* NA SA (*Panthera leo* *atrox) NA SA	*Panthera* Af As NA SA, lion, jaguar, tiger
	†*Miracinonyx* NA	*Acinonyx* Af As, cheetah
Family Hyaenidae		*Hyaena* Af As, striped hyena

TABLE 3 *continued*

Higher category	Extinct Genus	Living Genus
	Crocuta As	*Crocuta* Af, spotted hyena
Family Canidae	(*Canis* *dirus) NA SA	*Canis lupus* NA As, wolf
Family Ursidae	†*Theriodictis* SA	
	†*Arctodus* NA SA	*Ailuropoda* As, giant panda
	†*Tremarctos* NA	*Tremarctos* SA, spectacled bear
		Ursus NA Af As, brown, grizzly, black bears
		Melursus As, sloth bear
		Helarctos As, Malayan sun bear
		Thalarctos As Eu NA Arctic O., polar bear

Order Primates

Family Lemuridae	†*Megaladapis* Ma, 35–75 kg	
†Family Archaeolemuridae	†*Archaeolemur* Ma, 15–25 kg	
	†*Hadropithecus* Ma, 28kg	
†Family Palaeopropithecidae	†*Babotokia* Ma, 15 kg	
	†*Archaeoindris* Ma, 200 kg	
	†*Palaeopropithecus* Ma, 45–55 kg	
Family Indridae	†*Mesopropothecus* Ma, 10 kg	
Family Cercopithecidae		*Theropithecus* Af, gelada baboon
		Papio Af, baboons, mandrill, drill
Family Hominidae		*Pongo* As, orangutan, mia
		Pan Af, chimpanzee, bonobo
		Gorilla Af, gorilla

TABLE 3 *continued*

Higher category	Extinct Genus	Living Genus
		Homo humans, worldwide
Order Tubulidentata		
Family Orycteropodidae		*Orycteropus* Af, aardvark
Order Artiodactyla		
Family Suidae		*Sus* Af As, pigs, wild boars
Potamochoerus		
		Af Ma, bush pig, river hog
		Hylochoerus Af, forest hog
	(Phacochoerus) As [Israel]	*Phacochoerus* Af, warthog
		Babyrousa E. Indies, babirusa
Family Tayassuidae	†*Mylohyus* NA †*Platygonus* NA SA †*Brasilochoerus* SA	
Family Hippopotamidae	*(Hippopotamus)* As, MI, Ma	*Hippopotamus* Af, hippopotamus, dwarf hippopotamus
	(Hexaprotodon) As, Java	*Hexaprotodon* Af, pigmy hippopotamus
Family Camelidae	†*Hemiauchenia* NA	*Camelus* As, camel, dromedary, Bactrian camel
	†*Camelops* NA	*Lama* SA, lama, alpaca, guanaco
	†*Eulamaops* SA †*Paleolama* NA SA	*Vicugna* SA, vicuña
Family Antilocapridae	†*Tetrameryx* NA	*Antilocapra* NA, pronghorn, American antelope
Family Cervidae	†*Torontoceros* NA	*Cervus* Af As Eu NA, red deer, stag, elk, sika

TABLE 3 *continued*

Higher category	Extinct Genus	Living Genus
	†*Metacervulus* As	*Axis* As EI, spotted deer, chital, axis deer
	†*Sinomegaceros* As	*Elaphurus* As, mi-lu, Pere David's deer
	†*Candiacervus* As (Crete)	*Dama* Af As Eu, fallow deer
	†*Megaloceros* As Eu	*Capreolus* As Eu, roe deer, roebuck
	†*Megaloceroides* Af	*Alces* As Eu NA, moose, European elk
	†*Cervalces* As Eu NA	*Odocoileus* NA SA, white-tail, mule deer
	†*Bretzia* NA	*Blastocerus* SA marsh deer, swamp deer
	†*Antifer* SA	*Ozotoceros* SA
	†*Charitoceros* SA	*Rangifer* As Eu NA, caribou, reindeer
	†*Agalmaceros* SA	*Hippocamelus* SA, Andean deer, huemul
	†*Navahoceros* NA	
Family Giraffidae	†*Sivatherium* Af, As	*Okapia* Af, okapi
		Giraffa Af, giraffe
Family Bovidae	†*Spirocerus* As	*Gazella* Af As, gazelles, chinkara, korin, goa
	†*Pelorovis* Af	*Antilope* As, blackbuck, Indian antelope
	†*Myotragus* Balearics, Holocene	*Antidorcas* Af, springbuck
	†*Bootherium* NA	*Pantholops* As, chiru, Tibetan antelope
	†*Euceratherium* NA	*Litocranius* Af, gerenuk, giraffe gazelle

TABLE 3 *continued*

Higher category	Extinct Genus	Living Genus
	(Saiga) NA	*Saiga* As, saiga
	†*Parmularius* Af	*Capra* Af As Eu, goats, ibexes, tur, markhor
	†*Rhynotragus* Af	*Ovis* Af As Eu NA, sheep, mouflon, bighorn
	†*Rusingoryx* Af	*Hemitragus* As, tahr
	(Ovibos) As	*Ammotragus* Af, Barbary sheep, aoudad, arui
		Rupicapra As, chamois
		Oreamnos NA, Rocky Mountain goat
		Pseudois As, blue sheep, bharal
		Budorcas As, takin
		Ovibos NA, musk ox
		Capricornis As, serow
		Pseudoryx As, sao la, Vu Quang ox
		Boselaphus As, nilgai, blue bull
		Syncerus Af, African buffalo
		Bos As, cattle, aurochs, yak, gaur, kouprey
		Bubalus As, water buffalo, Asiatic buffalo
		Bison As Eu NA, bison (American buffalo), wisent
		Anoa East Indies, dwarf buffalo, tamarau

TABLE 3 *continued*

Higher category	Extinct Genus	Living Genus
		Tragelaphus Af, kudu, sitatunga, bushbuck
		Taurotragus Af, eland
		Boocercus Af, bongo
		Redunca Af, reedbuck
		Kobus Af, lechwe, kob, puku
		Hippotragus Af, roan, sable, blaauwbok
		Oryx Af As, gemsbok, oryx
		Addax Af, addax
		Alcelaphus Af As, hartebeest, kongoni, tora
		Connochaetes Af, wildebeest, gnu
		Damaliscus Af, blesbok, bontebok, topi, tiang
		Sigmocerus Af, Lichtenstein's hartebeest
		Aepyceros Af, impala
		Cephalophus Af, forest duiker, blue duiker
†Order Litopterna †Family Macraucheniidae	†*Macrauchenia* SA †*Xenorhinotherium* SA	
†Order Notoungulata		
†Family Toxodontidae	†*Toxodon* SA †*Mixotoxodon* SA NA	

TABLE 3 *continued*

Higher category	Extinct Genus	Living Genus
Order Perissodactyla		
Family Equidae	*(Equus)* NA SA	*Equus* Af As, horse, donkey, tarpan, onager
	†*Hippidion* SA	
	†*Onohippidion* SA	
Family Rhinocerotidae	†*Stephanorhinus* Eu Af	*Dicerorhinus* As, Sumatran (hairy) rhinoceros
	†*Coelodonta* As Eu	*Rhinceros* As, one-horned (Indian) rhino
		Ceratotherium Af, African white rhinoceros
		Diceros Af, African black rhinoceros
Family Tapiridae	†*Megatapirus* As	*Tapirus* NA SA As, tapir, danta, huagra
Parvorder Proboscidea		
†Family Mammutidac	†*Mammut* NA	
†Family Gomphotheriidae	†*Cuvieronius* NA SA	
	†*Haplomastodon* SA	
	†*Stegomastodon* SA	
Family Elephantidae		*Elephas* As, Asian elephant
	†*Mammuthus* NA, As	*Loxodonta* Af, African elephant
	†*Stegodon* As, Holocene? China	
Total genera, large size	Au = 14; NA = 31 (2); SA = 47; NA + SA = 12; Ma = 3 + 4 small; Af = 8; As = 12; Af + As = 15. Total = 102 (+ 4, small, Ma)	Au = 2; NA = [13] 14; SA = 12; NA + SA = [20] 22 or 23; Af = 53; As = 40; Af + As = 77; Ma = 1. Total = 92

NOTE: NA = North America (excluding West Indies); SA = South America; Ma = Madagascar; Af = Africa; As = Asia, including Europe; Eu = Europe; Au = Australia; NG = New Guinea; MI = Mediterranean islands.
†Extinct genus.
*Extinct species or subspecies.
SOURCE: Derived from McKenna and Bell 1997 (© Columbia University Press).

lions and spotted hyenas of Africa, once lived in cold climates with the woolly mammoths. The extinction of a primate species, the Neanderthals, by anatomically modern people in western Europe around 50,000 years ago is reviewed by Richard Klein (1999). This was close to the time of the extinctions of hippo and temperate rhino in Eurasia (Stuart 1999). In near time, tropical Asia, like Africa, suffered minor losses, less than 10 percent of its megafauna. Unlike those in the Americas, the losses were not spontaneous but more gradual.

None of this means that Africa and Asia can be downplayed in our explorations of what happened to the Pleistocene world. Pleistocene extinctions are an important piece of evidence in assessing the causes of near-time extinctions worldwide. Plio-Pleistocene losses of Afro-Asian elephants, suids, large carnivores, and primates may reflect hominid competition or predation (Klein 1999; Martin 1966, 1967a, 1984; Surovell and Brattingham 2005). Of all the continents, only Antarctica, home of pinniped rookeries and penguin breeding colonies, but never of prehistoric humans living off the land, escaped near-time losses.

AUSTRALASIA

Australasia includes Australia and New Guinea. Australia's endemic mammals belong mainly to the order Marsupialia (the pouched mammals). Fewer large-mammal genera went extinct in Australia than in North America. However, one might argue that since it is smaller in area than North America, has a drier climate with poor soils (no vast rich loess deposits along a major interior drainage like the Mississippi), Australia deserves a handicap. In fact, if we define large mammals in Australia, which do not exceed 1,000 kilograms (2,200 pounds), as ranging down to 15 kilograms (about 30 pounds) in adult body weight, and compare them with those North American large mammals that range from 3,000 kilograms (6,600 pounds) down to 40 kilograms (roughly 90 pounds), and compare the extinction of species rather than genera, the records are comparable. Peter Murray (1991) lists 46 extinct species between 40 and 3,000 kilograms for North America north of Mexico and 41 extinct species between 15 and 1,000 kilograms for Australia (see figure 5). The number of taxa lost to extinction is similar; the mass of the species lost is greater in America.

Australasia saw the near-time extinction of *Thylacoleo,* the only genus of large mammalian carnivore that it had to lose. The best-known species of this genus, *T. carnifex,* is sometimes called the "marsupial lion."

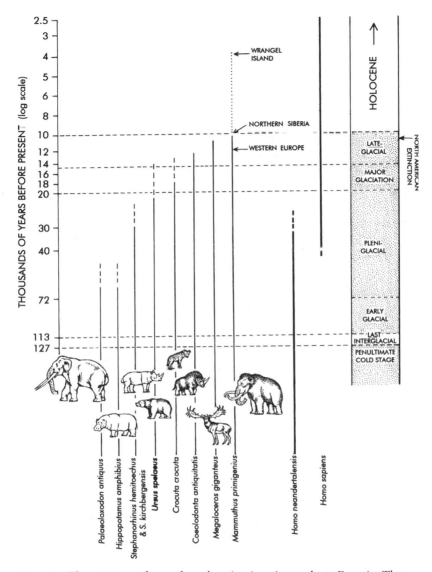

Figure 3. The patterns of megafaunal extinctions in northern Eurasia. The timelines estimate the latest survival of selected large mammals in Europe and Siberia, including their relation to environmental changes and to replacement of Neanderthals by modern humans. The inferred extinction dates are staggered compared with the pulse of extinctions in North America at 11,000 C14 years ago. First to disappear are straight-tusked elephant and hippo; the last are the giant deer and woolly mammoth. The latter persisted on Wrangel Island until after 4000 C14 years ago. It is clear that extinctions do not correlate with episodes of climatic change. They suggest that extinctions of temperate species (elephant and hippo) preceded those better adapted to high latitudes and colder climates as modern humans became better adapted to higher latitudes. Reprinted from MacPhee, ed., 1999. Used with kind permission of Springer Science and Business Media.

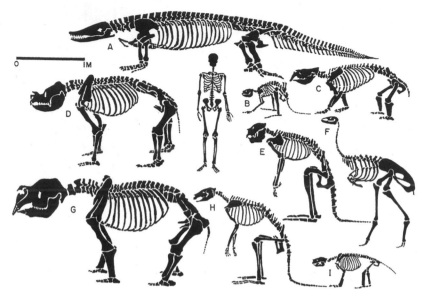

Figure 4. Extinct Australian megafauna, scaled. A. *Megalania* (giant varanid lizard); B. *Simosthenurus;* C. *Phascolonus;* D. *Zygomaturus;* E. *Procoptodon;* F. *Genyornis;* G. *Diprotodon;* H. *Macropus titan;* I. *Thylacoleo.* From Murray 1991, in P.V. Rich et al., *Vertebrate paleontology of Australasia.*

It weighed up to 130 kilograms (about 285 pounds), intermediate in size between an African lion and a leopard. Although too old to date by radiocarbon, a remarkable number of well-preserved *Thylacoleo* carcasses have been discovered in sinkholes in the Nularbor Plain, a vast region of flat-lying karst limestone in south-central Australia. The carnivorous nature of *Thylacoleo* is suggested by its prominent incisors and large, bladelike premolars. It also had a large, hooked claw on each thumb, and, like us, it could move its thumbs independently of its other fingers, a useful feature if it was arboreal, as some paleontologists suspect was the case.

In Australia the predator-scavenger niche may have been partly filled by reptiles, as it is today. *Megalania,* a giant varanid or monitor lizard 5 to 9 meters (16 to 30 feet) in length, may have attained 880 kilograms, or 2,000 pounds (Molnar 2004). This is almost ten times the weight of its close relative the "ora" or Komodo dragon, the leopard, and the extinct *Thylacoleo,* according to recent estimates by Stephen Wroe (Wroe and others 2003). In weight, *Megalania* may have matched America's short-faced bear, *Arctodus,* as a giant carnivore. However, its reptilian physiology would have tolerated longer periods of starvation. Some sus-

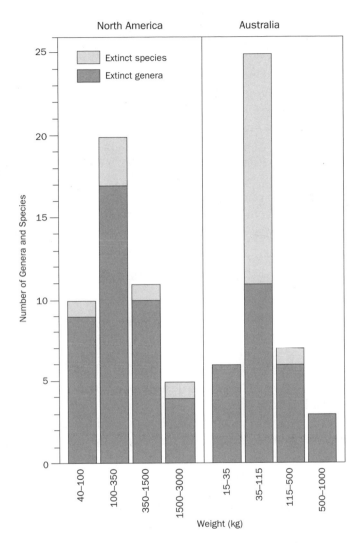

Figure 5. Comparison of North American and Australian megafaunal extinctions. From Murray 1991, in P.V. Rich et al., *Vertebrate paleontology of Australasia.*

pect that scarcity of *Megalania* fossils may be attributable to a cannibalistic habit seen in living Komodo dragons (Murray and Vickers-Rich 2004). Australia also had two terrestrial crocodiles, a giant python, and (although not a carnivore) a giant "horned" land turtle.

The largest marsupials of Australia and New Guinea, the diprotodons, were roughly the shape and weight of a small rhinoceros. Early in near time these, too, vanished from the fossil record. They have no living relatives closer than the largest living marsupials, the red and gray kangaroos, which are in a different family. The stocky diprotodons were not designed to bound along like kangaroos.

Australia lost a number of its kangaroos. There had been a proliferation of kangaroo species in two genera, *Sthenurus* (now extinct) and *Macropus* (the genus harboring multiple living species, plus many fossil species). While the living large kangaroos are grazers, the extinct kangaroos included browsing species. As with other Australian marsupials, the extinct kangaroo species appear to have been less swift, not as well designed as the surviving large kangaroos to escape cursorial predators (those adapted for running). Australian paleontologists such as Mike Archer, Peter Murray, and Tom Rich offer reconstructions (see figure 4).

THE AMERICAS

A large group of terrestrial mammals unique to the Americas and rich in vanished giants is Xenarthra ("strange joint"), formerly called Edentata ("toothless"), although only the anteaters have a reduced dentition. Originating in South America, Xenarthra includes such living and extinct representatives as armadillos, giant armadillos, and glyptodonts (order Cingulata, "banded"), as well as anteaters, ground sloths, and tree sloths (order Pilosa, "hairy").

Except for wide variation in scale, the extinct cingulates generally resembled the nine-banded armadillo *(Dasypus novemcinctus)*, the bony-shelled creature is found in the southern United States today. The extinct species include the "beautiful armadillo" *(Dasypus bellus)*, perhaps twice the length of a living nine-banded armadillo and heavy enough to be considered large as defined here. Much larger was the pampathere, *Holmesina*, about 1 meter (40 inches) high and 2 meters (80 inches) long. It weighed 180 kilograms (400 pounds).

As for the glyptodonts, the largest species, *Glyptodon clavipes* of South America, weighed over 1,800 kilograms (4,000 pounds)—by an order of magnitude the largest "giant armadillo" of all (Lyons, Smith, and Brown 2004). North America had just one species of glyptodont in near time, which ranged south from Florida (and also central Sonora, at the same latitude as Florida); South America harbored many more species of giant edentates, both glyptodonts and ground sloths. With the solitary exception of the recently described *Pachyarmatherium*, only known earlier in the Pleistocene, all eight glyptodont genera recorded

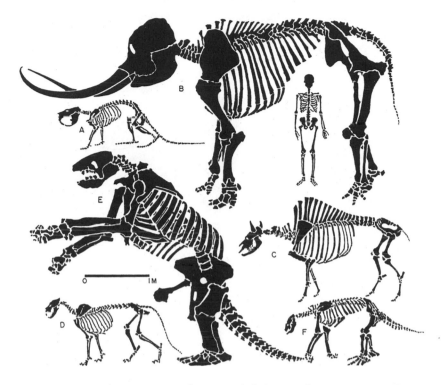

Figure 6. Extinct American megafauna, scaled. A. giant beaver; B. mastodon; C. bison; D. saber-toothed cat; E. giant ground sloth; F. Shasta ground sloth. From Murray 1991, in P.V. Rich et al., *Vertebrate paleontology of Australasia.*

in the last four million years vanished in the late Pleistocene, very likely in near time.

Because the ground sloths play an important role in this book, and given their importance in focusing Thomas Jefferson's interest over two centuries ago on the unknown large animals of America, their record is of more than ordinary interest.

I consider the ground sloths to be the hallmark, the defining group of mammals for the Americas. Ground sloths ranged from Alaska to Patagonia, including the West Indies. Lyons, Smith, and Brown (2004) list 14 extinct species in 12 extinct genera from South America; 10 of the extinct species exceeded 450 kilograms (1,000 pounds) adult body weight. As the hallmark of South America I nominate the largest ground sloth, *Megatherium americanum,* tipping the scales at over 6,000 kilograms (13,000 pounds), matching or exceeding the mass of a large bull Asian elephant. The largest giant ground sloth in North America,

Figure 7. Extinct North American megafauna, scaled: a. giant armadillo; b. giant ground sloth; c. Columbian mammoth; d. mylodon ground sloth; e. mastodon; f. Shasta ground sloth; g. glyptodont; h. camel; i. tapir; j. giant peccary; k. capybara; l. saber-toothed cat; m. American lion; n. horse; o. woodland musk ox; p. Harrington's mountain goat; q. antilocaprid. Reprinted from Stuart 1991. Used with permission of Cambridge University Press.

Eremotherium rusconii, is estimated to have weighed a little more than half as much.

In the late Quaternary, until their extinction, the ground sloths, of the order Xenarthra, comprised four genera in North America north of Mexico and twelve in South America (see table 4). These animals are a bit harder to envision than the armadillos. Their closest living relative, South America's giant anteater *(Myrmecophaga)*, at around 20 to 40 kilograms (40 to 85 pounds), is the largest survivor of this stunning lineage. The giant anteaters walk on their knuckles to protect the claws of their forefeet, and the giant ground sloths may have done the same. Like the anteaters, they may also have carried their young on their backs. However, they did not have the anteaters' long snout, very long tongue, reduced dentition, and specialized digestive tract, all designed to mop up and process termites or ants. The ground sloths had teeth and a digestive tract resembling those of a tree sloth, designed to process foliage.

Tree sloths spend most of their time suspended in tropical trees. Ground sloths may not have been quite as languid and slow moving, but their bulk, robust bones, and long, unretractable claws must have made it impossible for them to move quickly or to climb very well, and thus difficult for them to escape predators. Extrapolating from the fact that the related anteaters, tree sloths, and armadillos "are chiefly solitary but may form small, loose associations," Ronald Nowak (1999) writes that the ground sloths were probably solitary except in the breeding season. Thus, unlike elephants, camels, horses, bison, and the extinct relatives of those groups, they not only were not fleet of foot but also lacked the protection of a herd. They must nevertheless have succeeded somehow in defending themselves against nonhuman carnivores, perhaps by sitting up on their haunches, propped by their tails, and, like giant anteaters, using the curving claws on their long front legs to rip attackers. This defense would have had little effect against stones or spears flung from a distance by human hunters.

The ground sloths were the largest of the xenarthrans. Those that are estimated to have exceeded 450 kilograms (1,000 pounds) in weight were the true giant ground sloths. They would include *Eremotherium rusconii,* known from the southeastern United States south into South America; *Paramegatherium* spp.; and *Lestodon armatus* (Lyons, Smith, and Brown 2004). As mentioned above, the most massive of all was South America's *Megatherium americanum.* It is thought that *Megatherium* (which had four digits and three well-developed front claws) and *Eremotherium* (with three fully developed digits and two claws) were able to reach tree

TABLE 4 *Near-Time Genera of Extinct Ground Sloths*

Family Mylodontidae

1. *Mylodon* Owen, 1839, SA
2. *Glossotherium* Owen, 1839, NA, SA
3. *Paramylodon* Brown, 1903, NA
4. *Oreomylodon* Hoffstetter, 1949, SA
5. *Mylodonopsis* Cartelle, 1991, SA
6. *Lestodon* Gervais, 1855, SA
7. *Lestodontidion* Roselli, SA (Uruguay)
8. *Scelidotherium* Owen, 1839, SA
9. *Catonyx* Ameghino, 1891, SA

Family Megatheriidae

10. *Megatherium* G. Cuvier, 1796, SA
11. *Eremotherium* Spillmann, 1948, NA, SA

Family Nothrotheriidae

12. *Ocnopus* Reinhardt, 1875, SA
13. *Prezfontanatherium* Roselli, 1976, SA (Uruguay)
14. *Nothropus* Burmeister, 1882, SA
15. *Nothrotherium* Lydekker, 1889, SA
16. *Nothrotheriops* Hoffstetter, 1954, NA

Family Megalononychidae

17. *Diodomus* Ameghino, 1885, SA
18. *Paulocnus* Hooijer, 1962, Curaçao
19. *Synocnus* Paula Couto, 1967, Hispaniola
20. *Valgipes* Gervais, 1874, SA (Brazil)
21. *Megalonyx* Harlan, 1825, NA, SA (Colombia)
22. *Megalocnus* Leidy, 1868, Cuba
23. *Neocnus* Arredondo, 1961, Cuba, Hispaniola
    (*Microcnus* Matthew, 193l, and *Cubanocnus* Kretzoi, 1968)
24. *Parocnus* Miller, 1929, Cuba, Hispaniola
25. *Miocnus* Matthew, 1931, Cuba
26. *Acratocnus* Anthony 1916, Cuba, Puerto Rico, Hispaniola
27. *Xenocnus* Paula Couto, 1980, Brazil

NOTE: NA = North America, SA = South America. Taxonomy follows McKenna and Bell, revised by Greg McDonald and Ross MacPhee. Eighteen genera of ground sloths are confined to the Americas, from Alaska to Patagonia, with a secondary center of radiation of seven genera of dwarf (under 50 kg) animals in the Caribbean (Greater Antilles plus Curaçao). Living tree sloths—*Choelepus*, the two-toed sloth, and *Bradypus*, the three-toed sloth—occur in tropical America.

branches at least as high as those browsed by giraffes. The molar cusps of these monstrous beasts resembled giant pinking shears, suggesting that they were superbly adapted for browsing. The Uruguayan paleontologist Fariña (1996) proposed that in the absence of more suitable carnivores, *Megatherium* could have also been a facultative carnivore or scavenger, and this behavior is depicted in the Discovery Channel's 2001 television special *Walking with Prehistoric Beasts*. But *Megatherium* lacks carnassials, the shearing teeth that are a hallmark of most carnivores.

Recently some have speculated that giant ground sloths used their claws to prey on their relatives, the armored glyptodonts, an idea I find as fanciful as Peter Lund's proposal of 150 years ago that giant ground sloths, like living tree sloths, would clamber about in the giant trees of the tropics (Wallace 2004, 171). To suggest that the ground sloths were predators or even scavengers is a reach. The ground sloth dung deposits I have studied in the Grand Canyon and in caves in Nevada, New Mexico, and West Texas harbor no more traces of bone than I have seen in horse, mule, and burro manure, namely none.

Two medium-sized genera were America's *Paramylodon,* "near molar tooth" (sometimes known as *Glossotherium,* "tongue beast"), and North America's *Megalonyx* ("great claw"). The mylodons were rhinoceros-sized and possessed a dermal armor formed of closely spaced, pea-sized bones embedded in their thick hides, which presumably helped shield them from predators.

The megalonychids include the first ground sloth discovered in North America, Jefferson's ground sloth *(Megalonyx jeffersonii).* Thomas Jefferson informally named the genus when in 1799 he was sent bones, including claws, from a cave in what is now West Virginia. Because Jefferson, along with many others of his time, believed in a "great chain of being" in which all species were interdependent (if one link was broken, the chain would fail), he did not believe in extinction. *Megalonyx* must be alive! Jefferson originally interpreted the claws as coming from a large cat, and because they were three times the length of an African lion's claws, he assumed that the animal was three times the size of that beast—and hoped Lewis and Clark might catch sight of it. Later he came to realize that they were from an animal similar to the *Megatherium* described by the French paleontologist Georges Cuvier. Jefferson also believed that mammoths roamed unknown parts of the West. These ideas were not entirely fabulous. Subsequent fossil discoveries revealed that until the end of the last ice age, ground sloths and various proboscideans, including mammoths, had indeed been native to America. In addition, the fossil record would show that there had been lions in America contemporary with the extinct ground sloths and elephants of near time. Despite having filled Monticello with the bones of late-Pleistocene mastodons and other large mammals, Jefferson is still regarded by some paleontologists as unworthy of being called the "father of paleontology." His interest in this aspect of American prehistory is nevertheless prescient of the theme I develop here, that an understanding of what the New World once harbored is far from irrelevant; indeed, it is crucial to how future management might be envisioned and designed. Jefferson focused on fossils of large extinct animals that only now are beginning to gain their full measure of appreciation. If he was wrong in believing that America might harbor a lion three times larger than its close relatives in the Old World, America nevertheless harbored until not long ago not only lions, but also many other large predators eclipsing the ones we know historically and from the last hundred centuries. I think Jefferson's curiosity would have been aroused. What is the full implication of America's extinct lions, mammoths, and ground sloths?

For those of us in the Southwest, the North American ground sloth most likely to draw attention, with a reconstructed carcass on exhibit at Kartchner Caverns near Benson, Arizona, is the Shasta ground sloth, *Nothrotheriops shastensis*. It was about the size of a black bear, with a body mass of 135 to 545 kilograms (300 to 1,200 pounds). This made it the smallest continental North American ground sloth, at less than one-tenth the weight of the true giant ground sloths, which some taxonomists place in the same family. It was more than an order of magnitude more massive than the tree sloths, which weigh up to 10 kilograms (roughly 20 pounds). Its dung and soft parts are found occasionally in dry caves that also harbor preserved plant remains gathered by packrats, revealing details about the habitat as well as the diet of the Shasta ground sloth. In its heyday, the animal ranged from northern California and the Texas Panhandle south into the Sierra Madre Oriental of northern Mexico.

Five genera of elephants in three families (all in the order Proboscidea) occurred in the Quaternary of North America. Three of the genera—*Mammuthus* (mammoths), *Mammut* (mastodons), and *Cuvieronius* (gomphotheres)—survived into near time, close to 13,000 years ago. Both mammoths and mastodons stood roughly 3 meters (8 to 10 feet) high at the shoulder, weighed 6,000 kilograms (9,000 to 13,000 pounds), and had long, curving tusks and in some cases long hair. Mammoths are the extinct genus most likely to be found processed or butchered at prehistoric kill sites in North America. The largest terrestrial mammal in the hemisphere was the imperial mammoth, *Mammuthus imperator,* at 10,000 kilograms (22,000 pounds). Another elephant, *Stegomastodon superbus,* tipped the scales at 7,580 kilograms (16,700 pounds) and was South America's only mammal more massive than *Megatherium.*

Three or more species of elephantlike gomphotheres (Gomphotheriidae) lived in South America until the end of the Pleistocene. The genus *Cuvieronius* ranged north of Mexico to overlap in Florida with the mammoths and mastodons. Bones dating from the early Quaternary have been found in Arizona. The gomphotheres had relatively straight tusks supporting a strip of enamel, rather than tusks entirely enclosed in enamel. They had elongated lower jaws. Some species, though probably not the American ones, had lower as well as upper tusks. The Central American gomphotheres may have eaten (and hence dispersed) the large fruits of wild avocado, oil palm, and guanacaste trees. Ecologist Dan Janzen believes that they played a major role in the dispersal of the seeds of palatable fruits, just as today's elephants do in Africa. After the extinctions, humans—and later, domestic livestock—may have unknowingly substituted for the gomphotheres in dispersing such seeds (Janzen and Martin 1982; Barlow 2000; Dudley 2000).

The South American notoungulate and litoptern orders, which together spun off 19 families in over 60 million years (McKenna and Bell 1997), are the least familiar of America's extinct big herbivores of near time. Four genera survived to the end of the ice age. They were diverse groups, and envisioning them is a challenge. Notoungulates may have resembled rhinos with a mouth full of buck teeth, while litopterns resembled horses, camels, or chalicotheres (also extinct), large horse-like animals that had claws rather than hooves. The litoptern *Macrauchenia* looked vaguely like a large llama, with a much longer neck and an elongated, bootlike nose resembling a short trunk, with the nasal apertures on top of its head rather than terminal. *Toxodon,* a notoungulate, and *Macrauchenia* made rare appearances on the Discovery Channel's *Walking with Prehistoric Beasts.*

The order Carnivora comprises bears, cats, hyenas, mongooses, raccoons, seals, skunks, viverrids (civet cats and their kin), walruses, weasels, and wolves. Six genera of large carnivores disappeared considerably before near time; five more disappeared within near time, and four have survived (see table 2). Besides living wolves, bears, jaguars, and pumas, a zoological garden stocked with the American carnivores of near time would include the dire wolf, *Canis dirus;* two bears, *Tremarctos flori-danus* and *Arctodus;* a scimitar cat and a saber-toothed cat, *Homotherium* and *Smilodon;* a lion, *Panthera leo atrox;* and a cheetah, *Miracinonyx.*

At least some of these carnivores, such as the cheetah, dire wolf, and lion, are very close to living species. The dire wolf may have been more robust and less cursorial than living timber wolves. Hundreds of its skulls from the tar pits at Rancho La Brea decorate a wall panel at the Page Museum in Hancock Park, Los Angeles. Most paleontologists interpret the abundance of dire wolf and saber-toothed cat remains at Rancho La Brea as reflecting the fate of opportunistic scavengers who came to feast on whatever was trapped in the asphalt and became trapped themselves.

Judged by measurements of its carnassial teeth, the short-faced bear, *Arctodus simus,* was larger than any living bear, even the polar bear. The Florida cave bear, *Tremarctos floridanus,* while smaller than *Arctodus,* was much larger than its living relative, the spectacled bear, a vegetarian that survives in South America.

Two medium-sized North American carnivores that went extinct in near time were the dhole *(Cuon alpinus)* and a short-faced skunk *(Brachyprotoma).* The skunk may have suffered extinction when the reduction in megafauna deprived it of sufficient carrion. The dhole, a type of wild dog, survives in Asia, where it hunts in packs. Asian dholes weigh up to 21 kilograms (about 45 pounds) and are reddish in color with black, bushy tails; in appearance they would appeal to any dog lover. They emit distinctive whistling calls to reassemble separated pack members.

Of the odd-toed ungulates (order Perissodactyla), only horses and tapirs survived into near time. New World equids declined from 12 genera over 10 million years ago (S. Webb 1984) to only three genera in near time: *Equus, Hippidion,* and *Onohippidium.* Apart from these three, American horse extinctions long predate the arrival of humans. The genus *Equus* remained highly variable, with many species in the Americas, until near time. The explosive spread of free-ranging horses into grasslands of both North and South America following their reintroduction by the Spanish suggests a return of the native. Tapirs, too, went extinct in the United States in the late Quaternary, having previously lived not only in Florida but also in California, Kansas, and in Arizona, from the Sonoran lowlands to an elevation of over 6,000 feet in the Colorado Plateau. Three species of tapir survive from southern Mexico through Central and into South America, from tropical lowlands into the cold, wet, high elevations of the cordillera.

In North America the large living genera of the order Artiodactyla (the even-toed ungulates) include musk ox *(Ovibos)*, bison *(Bison)*, moose *(Alces)*, wapiti or elk *(Cervus)*, caribou *(Rangifer)*, deer *(Odocoileus)*, bighorn *(Ovis)*, mountain goat *(Oreamnos)*, and pronghorn *(Antilocapra)*. There were twice as many species of North American artiodactyls before the extinctions of near time.

To the surprise of those who do not know the fossil record, these included three genera of camels and llamas (family Camelidae), all now extinct. *Camelops* was the size of a dromedary, with longer legs and steeply sloping hindquarters. There were various species of llama in the genera *Paleolama* and *Hemiauchenia*. The latter is considered more closely related to the living South American genus, *Lama*. Some paleontologists find it ironic that for tens of millions of years both camelids and equids evolved in North America, only to migrate into and survive in Eurasia and South America, while they vanished in near time in their evolutionary heartland (Hulbert 2001).

Two now-extinct genera of peccary (piglike animals larger than the three living species) also roamed the continental United States. Both stood about 75 centimeters (30 inches) tall at the shoulder and had downward-pointing tusks. Fossils of the long-nosed peccary, *Mylohyus nasutus*, are often found in caves. Its low-crowned cheek teeth suggest that it ate fruits, nuts, and succulent vegetation. *Platygonus*, the flat-headed peccary, with its higher-crowned teeth, probably ate more cacti and other coarse vegetation (Hulbert 2001). Represented at 116 localities east of the continental divide and one-tenth as many to the west, *Platygonus compressus* may have been the most common of the medium-sized extinct mammals. It probably lived in herds. Attaining 50 kilograms (110 pounds), the weight of an Old World boar, *Platygonus* was considerably heavier than the three living species, the collared and white-lipped peccaries and the relatively recently discovered Chacoan peccary, *Catagonus wagneri*, a rare South American species that can attain 40 kilograms (almost 90 pounds).

Extinct cervids (members of the deer family) include *Cervalces*, the stag moose, commonly found in eastern Quaternary faunas; *Torontoceros*, a recently described and poorly known deer; *Bretzia*, a very large deer dating from the early and evidently also the late Quaternary; and *Navahoceros*, the mountain deer, thought to be related to living Andean deer.

Extinct bovids (members of the cattle family) include the woodland musk ox, which has mistakenly been given two generic names (*Bootherium* for the female, *Symbos* for the male; the former has priority), and the brush ox, *Euceratherium*. The extinct antilocaprids, *Capromeryx*,

Stockoceros, and *Tetrameryx,* all had divided (V-shaped) horns, rather than *Antilocapra*'s pronghorn. *Capromeryx* and *Stockoceros* were also smaller than the pronghorn, an exception to the rule that smaller members of a lineage were more likely to survive extinction in near time.

The largest American rodent to go extinct in near time was the bear-sized giant beaver, *Castoroides.* This creature did not build dams like living beaver. It would have been a major drawing card in an imaginary pre-extinction zoo, rivaled by the similarly sized extinct capybara, *Neochoerus,* found fossil in Florida. Another capybara, *Hydrochoerus,* persists in tropical America and is a popular "giant rodent" in zoos.

In the last four million years the North American continent north of Mexico lost 26 genera of small rodents. What is of interest for our analysis is that all of these extinctions predated the Rancholabrean and near time (Martin and Steadman 1999, table 2). Extinctions of rodent species in near time occurred on oceanic islands but rarely on the continent. Arthur Harris has described two extinct species of packrats *(Neotoma),* also known as wood rats, from the arid West. Ethnographic data indicate that packrats were popular prey for hunter-gatherers in much of their range. Possibly the extinction of the two species described by Harris came at the hands of prehistoric foragers.

Of the two genera of North American sirenians known north of Mexico, *Trichechus,* the manatee (600 kilograms, or roughly 1,300 pounds), survives in the coastal waters of Florida. In the Quaternary, *Hydrodamalis stelleri,* Steller's sea cow (over an order of magnitude heavier than a manatee), disappeared from the coastal waters of California and Alaska, and a related species disappeared from Japan. A population that probably did not exceed 1,000 to 2,000 animals survived in the kelp beds of the undiscovered and uninhabited Commander Islands of the northwestern Pacific. By 1768, less than 30 years after the arrival of explorers and fur traders, the genus was biologically extinct. This is the only historical extinction of a megaherbivore, apparently by overkill.*

ISLANDS

Perhaps a dozen extinct species of dwarf megalonychid sloths weighing roughly 5 to 70 kilograms (10 to 150 pounds), considerably less than their fossil relatives from North, Central, and South America and in some

---

*I define megaherbivores as plant-eating mammals exceeding a metric ton, or 1,000 kilograms (2,200 pounds).

cases equal to living tree sloths, are known from the fossil record of islands in the Caribbean, especially from Cuba and Hispaniola (White and MacPhee 2001).

According to Jennifer White and Ross MacPhee, "All mainland megalonychids . . . are thought to have become extinct by the end of the Pleistocene; Antillean taxa held on until the middle-late Holocene (at least on some islands), but all were gone well before European arrival" (White and MacPhee 2001). Eight taxa occurred in Cuba, six on the island of Hispaniola, two genera in Puerto Rico and one each in Grenada and Curaçao. A few specimens have been radiocarbon dated; the early returns indicate that the dwarf species lasted thousands of years longer than their massive continental relatives, their extinction coincidental with human colonization of the West Indies. Recent dating of ground sloth extinction on the American mainland and in the West Indies offers a critical test of the overkill model; the sloths on Haiti outlast those of the mainland.

Madagascar, an island the size of Texas that remained far enough east of the coast of Africa to have evolved numerous endemic species, saw the loss of *Plesiorycteropus,* long thought to be a relative of the African aardvarks. Its relationship to the aardvark family (Orycteropidae) is so remote, however, that mammalogist Ross MacPhee of the American Museum has placed it in an order of its own (Bibymalagasia). The larger of two described species is estimated to have weighed up to 18 kilograms (40 pounds); the smaller weighed 6 to 10 kilograms (13 to 22 pounds). Both were apparently better adapted for climbing than African aardvarks.

Beginning about 2,400 years ago, the big losers were primates: 16 species (6 genera) of lemurs, all larger than 10 kilograms (22 pounds) and the largest, *Archaeoindris,* approximating a gorilla in size (see figure 8). Species of *Megaladapis,* sometimes called the "koala lemur," are known from the last full glacial. Madagascar also lost hippos, giant tortoises, and *Aepyornis,* the elephant bird, the largest flightless bird of near time. Its eggs had a fluid capacity exceeding a gallon. In some coastal regions, *Aepyornis* eggshells litter the ground like the wrack of clamshells. Shrinkage of large herbivore biomass is reflected in large spore counts of *Sporormiella,* which decline with the extinctions of hippos and other large vertebrates and recover with the introduction of cattle (Burney and others 2004). Based on two radiocarbon dates, David Burney's group believes *Hippopotamus* survived on the east coast until only two centuries ago.

Around 500 years ago, New Zealand saw the extinction of 10 (or fewer) species of moa, flightless birds ranging in size from kiwis to larger

NZ

Figure 8. Extinct genera of lemurs of Madagascar. One of the largest living lemurs, *Indri*, shown for scale. 1. *Palaeopropithecus*; 2. *Megaladapis*; 3. *Babakotia*; 4. *Archaeolemur*; 5. *Hadropithecus*; 6. *Archaeoindris*. Reprinted from Simons 1997.

than ostriches. Unlike ostriches and emus, the moas presumably moved with their heads down for easier travel through dense forests.

Hawaii saw the near-time extinctions of many birds, including a flightless gooselike duck, *Thambetochen,* and a flightless ibis, *Apteribis,* both named by Storrs Olson and Alexander Wetmore of the Smithsonian Institution. Other islands in the remote Pacific lost endemic parrots, pigeons, doves, megapodes or bush turkeys *(Megapodius),* and especially flightless rails *(Gallirallus),* as well as large invertebrates such as land snails.

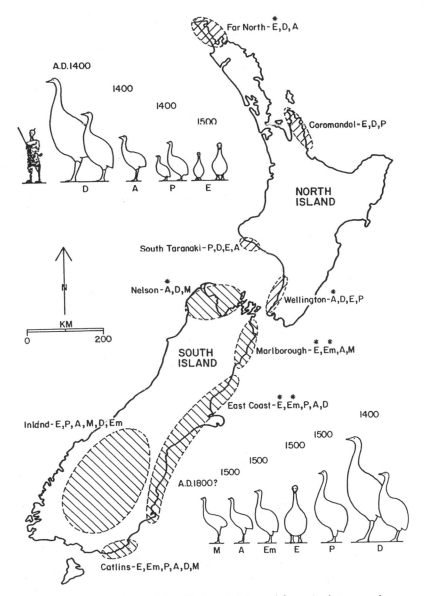

Figure 9. Moa extinction in New Zealand. Adapted from Anderson and McGlone 1992. Fossil moa are most numerous in the cross-hatched areas. A. *Anomlopteryx;* D. *Dinornis;* Em. *Emeus;* E. *Euryapteryx;* M. *Megalapteryx;* P. *Pachyornis.* According to Worthy and Holdaway 2002, the period of extinction was breathtakingly short.

Many rodents that reached oceanic islands such as the West Indies and the Mediterranean Islands and evolved into endemic genera did not survive human arrival along with the introduction of domestic rats *(Rattus)*. In some cases the rats appear to have colonized first.

In the case of insectivores, tiny fossorial mammals like moles and shrews, anthropogenic extinction in near time would not be expected. Exceptions can occur on oceanic islands, perhaps as the result of the introduction of human commensals such as rats. The list of near-time extinctions is astounding in itself. Even more astounding, I was to discover, was the apparent reason for them. That reason was taking shape for me in 1956 and is hotly debated 50 years later. I have spent many of those years searching for resolution, as the next chapters will indicate. I believe the evidence points overwhelmingly to one disturbing conclusion.

RADIOCARBON DATING
AND QUATERNARY EXTINCTIONS

The technique of radiocarbon dating, developed by Willard F. Libby in the 1940s and for which he was awarded a Nobel Prize in 1960, brought a revolutionary change to our understanding of many biological and physical events of near time. No other geochemical dating method is as powerful in aiding our understanding of dynamic changes over the past 50,000 years. Radiocarbon dating allows scientists to make the most accurate estimates possible for the timing of late-Quaternary extinctions. These estimates, in turn, are crucial in testing various models or ideas about the possible causes of near-time extinctions.

All living (organic) matter contains carbon, as does the Earth's atmosphere. Radioactive carbon, or radiocarbon (^{14}C), is a low-energy radioactive isotope, or variant, of carbon that is continuously being formed in the upper atmosphere by the action of cosmic radiation on nitrogen-14 (^{14}N). As do all radioactive molecules, the molecules of ^{14}C subsequently decay at a characteristic rate. The ratio of ^{14}C to nonradioactive carbon (^{12}C) in the atmosphere is very small.

As dynamic elements of nature's carbon cycle, living organisms maintain a small level of ^{14}C in their tissues, in the same ratio to ^{12}C as exists in the atmosphere. When an organism dies, the amount of radioactive carbon it contains is no longer replenished by exchange with the environment, and the amount of this radioactive carbon thus begins to diminish, decaying at a rate that reduces the number of ^{14}C atoms by half over the span of 5,730 years. This time span is called the "half-life" of ^{14}C. Radiocarbon dating measures the ratio of ^{14}C to ^{12}C in fossilized organic material and compares it to the ratio of ^{14}C to ^{12}C in the atmosphere. The difference is a function of the time since death. The older a specimen is, the less radioactivity it will retain. Measurements beyond 40,000 years (or seven half-lives) are possible but very difficult to obtain because of the heightened risk of older samples becoming contaminated (R. Taylor 1987).

Radiocarbon measurements are refined by comparing them to calendar years. For example, scientists can compare the ^{14}C measurements to a date determined by dendrochronology, or tree-ring dating. Or they can compare the measurements to the layers of annually laminated sediments, or to other organic material of known age, such as wood from tombs of the pharaohs.

Each age estimate is accompanied by an error range (plus or minus a certain number of years), which is partly a function of sample size and partly a function of the time it takes to make the measurement. In the past 25 years, the use of particle accelerators has made it possible to analyze (at somewhat higher cost than with previous methods) much smaller samples in much less time. This is an advantage because much less of the specimen needs to be sacrificed to obtain a date, and the risk of contamination is lower.

Radiocarbon dating assumes that the ratio of ^{14}C to ^{12}C in the atmosphere has remained constant over time. There have been fluctuations, however. One such fluctuation resulted from the combustion of fossil fuels such as coal and petroleum during the Industrial Revolution. This released large amounts of ^{12}C (nonradioactive carbon) into the atmosphere, reducing the $^{14}C : ^{12}C$ ratio. A much stronger shift in the opposite direction began with the testing of nuclear weapons; by 1963, nuclear tests had increased the atmospheric levels of ^{14}C by over 90 percent. In both cases, the balance of nonradioactive carbon to ^{14}C in the atmosphere was thrown off, impairing the accuracy of radiocarbon dating for specimens originating from the past two centuries. Trees that began growing in the mid-nineteenth century, or that predated but lived through the atomic age, are, according to radiocarbon dating, "too old" by 2 to 3 percent. With the elimination of atmospheric tests the production of ^{14}C is returning to normal background levels, those produced only by cosmic rays.

Early in the refinement of the technique of calibrating radiocarbon dates with tree-ring dates, investigators noticed another problem: minor but persistent departures in their graphs from the expected straight line. These "squiggles" or "wiggles" of a few decades, sometimes termed the de Vries effect, are attributable to changes in the strength of the Earth's geomagnetic field, which provides a shield against cosmic radiation.

Also early in the development of radiocarbon dating, geologists and ecologists recognized that carbonate samples taken from oceans or estuaries, and some lakes, would yield measurements that were too old. These water bodies are depleted of ^{14}C because the carbon they receive comes from streams that

The Geologic Time Frame
Calendar years before present (x 1000)

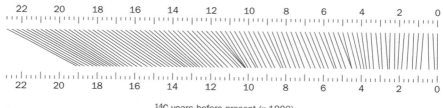

^{14}C years before present (x 1000)

Figure 10. Relation between radiocarbon and calendar, or calibrated, years, from about 22,000 calendar years ago to the present. Note how the discrepancy increases over time. Reprinted from Klein 1999. Courtesy the University of Chicago Press.

in turn receive their carbon from "old" rocks, which have relatively little ^{14}C. When the inorganic carbon from limestone or other carbonate rocks becomes incorporated metabolically into organic material like aquatic plants or animals in sediments, such organic material is much "too old" when dated by its radiocarbon content. Even tissues from *living* aquatic animals in such ^{14}C-depleted waters have been measured at thousands of years old.

In North America, mammoths and many other large mammals are thought to have gone extinct roughly two ^{14}C half-lives (11,460 years) ago. Since the amount of ^{14}C present in a specimen decreases by half during each half-life, this means that three-quarters of the ^{14}C originally found in the organic residues of the last living mammoths had disappeared between the time these mammoths died and the time when their remains were collected and submitted for dating. However, the dating of megafaunal extinctions around the world remains controversial.

Through the past several decades of active use of the radiocarbon dating method, there has been an understandable tension between those eager for results on the samples they have submitted and those pretreating the samples and evaluating them for the possibility of contamination. As a general rule, the best-preserved bones (and occasionally tissues) for dating come from frozen ground or from dry caves, especially those in arid regions. However, fossil bones from salt-impregnated sediments, or sediments impregnated with petroleum residues, may retain relatively large amounts of well-preserved connective tissue (collagen), which can be contaminated by carbon depleted of ^{14}C, thus throwing off radiocarbon measurement. Defensible dating requires removal of the petroleum residues from the specimens.

In continental North America it took at least two decades of measurements to discredit the belief that extinction of many large late-Quaternary mammals might have happened as late as 8,000 years ago or even more recently. Dates of this vintage continue to emerge from South America. Until recently, occasional Australian samples of extinct megafauna yielded measurements of less than 30,000 years ago, but these measurements have not been defended by replication at more than one laboratory. Of course, if many of the anomalously recent dates from continental North or South America or Australia prove to be defensible—that is, if they can be replicated by independent laboratories following established pretreatment protocol—a major readjustment in the theory of late-Quaternary overkill would be inevitable.

Another controversy centers on extinctions of relatively rare genera of large animals in North America. The few dates that have been posited from specimens of these animals are significantly older than 13,000 years. Those who believe that extinctions closely track human arrival discount these few older dates as insufficient for us to be confident that the genus in question became extinct significantly before the time of Clovis colonization, 13,000 years ago. Those who deny an anthropogenic role in the extinction process accept these dates at face value, suggesting that important megafaunal extinctions predated Clovis colonization.

The amount of ^{14}C in the atmosphere at any one time is a function of what is held in the Earth's main carbon reservoir, the oceans. As explained earlier, the oceans are ^{14}C-depleted (or ^{12}C-rich) in relation to the atmosphere. During some time periods, there is a "degassing" of carbon from the oceans (the reasons are not well understood, but one cause may be a change in temperature). When this happens, the atmosphere receives a higher percentage of ^{12}C (stable carbon) than is normal, diluting the amount of ^{14}C. Organisms living during such a time period thus also contain less ^{14}C than is normal, and that increases their apparent age when measured by radiocarbon dating. During the critical part of the time scale for extinctions in continental North and South America, from 10,000 to 13,000 ^{14}C years ago, just such dynamic changes were under way in the oceans. Thus there is less precision than one would wish in the radiocarbon dating method across the three millennia when Clovis hunters arrived and large mammals became extinct.

OVERVIEW OF OVERKILL

The real lesson of late Quaternary mass extinctions is
not that strange things happened in the past, a superficial
impression that is the only one this ongoing debate will leave
with the general public. Instead, Quaternary paleoecologists
should unite to publicize one of the greatest discoveries in
the history of paleontology: Long before the dawn of written
history, human impacts were responsible for a fantastically
destructive wave of extinctions around the globe. This message should be seen as a wake-up call instead of a mere omen
of disaster.

**John Alroy, "Putting North America's End-Pleistocene
Megafaunal Extinction in Context"**

Many forces that could trigger extinction are evident in Earth's history.
Proposed explanations for the near-time extinctions have included meteor strikes, climate change, nutrient shortages, and disease, among many
others. But for half a century, the explanation that has made the most
sense to me is what Richard Klein (1999, 564) calls the "ecological shock
of human arrival." Climatic change is always of interest but not crucial
in formulating explanations. As our species spread to various continents,
we wiped out their large mammals; as we progressed to oceanic islands,
we extinguished many mammals that were much smaller, and even more
birds, especially flightless species.

Based on the concept that animal populations could have sustained
some additional predation, but not as much as took place after human
arrival, this explanation has come to be known as "overkill." Some zo-

ologists, such as Ross MacPhee, agree with an anthropogenic explanation (one linked to human activity) but attribute the extinctions to "over-ill," the introduction by humans of predators, competitors, or disease vectors such as Pacific rats *(Rattus exulens)*. In any event, some aspect or aspects of human activity appear to have been the ultimate cause of these historic extinctions (Flannery and Schouten 2001).

The idea of overkill was not unknown when I began studying extinctions of near time. Yale's famous Quaternary geologist Richard Foster Flint brought up the possibility of human involvement in late-Quaternary extinctions (Flint 1971). Ed Deevey (and long before him, Darwin's contemporary, Sir Richard Owen) had suggested something similar in the case of New Zealand's moas. And by the 1950s, work by vertebrate paleontologist Claude Hibbard and by paleontologists in South America indicated that the large mammals of the Americas disappeared at a time when our species or its ancestors were present. Nevertheless, Hibbard and many other vertebrate paleontologists favored climatic change as the ultimate explanation for prehistoric extinction of terrestrial vertebrates in the Americas.

To test the hypothesis of human involvement, it would be vital to establish relatively exact dates for both the extinctions and the arrival of humans at the places the extinctions occurred. The geographic and chronological patterns of the extinctions should help reveal their cause. Identifying these patterns, however, posed a challenge. Our knowledge of extinct faunas had long been dependent on stratigraphy (study of the composition and order of the strata of the Earth's crust), chronology (age of the extinctions), and paleontology (the study of fossils). None of these allowed for dating on the scale required for our purposes. The breakthrough came with radiocarbon dating and the possibility of recovering age estimates of late-Quaternary extinctions accurate to within a century or less.

Willard Libby's new method of radiocarbon dating promised to allow human and nonhuman fossils from within roughly the last 50,000 years to be dated with precision, yielding age estimates in the last 20,000 years (dates with errors of at most a few hundred years). By establishing dates for the known fossils of any given species, we could make educated guesses about when that species went extinct and whether the extinction was sudden or gradual. For example, it was now possible to demonstrate that Wrangel Island, in the Arctic Ocean north of Siberia, sustained a population of woolly mammoths thousands of years after the species had gone extinct in the rest of its range. Radiocarbon dates tell us that woolly mammoths endured on Wrangel into the time of the pharaohs, roughly 4,000 years ago (Vartanyan, Garutt, and Sher 1992). In comparison, the extinction date for mammoths in western North America is now set at around 11,000 radiocarbon or 13,000 calendar years ago. In the early days of less reliable radiocarbon measurement, the extinction of mammoths and other large mammals was mistakenly thought to be more recent. Now we know it was, not only on Wrangel but also on St. Paul in the Pribilofs in the Bering Sea (Guthrie 2004). By the same token, dates on dwarf mammoths on San Miguel, off the California coast, agree in age with those from the mainland. These details suggest that Wrangel and the Pribilofs were temporary refugia, bypassed by the first Americans.

Thanks to radiocarbon and other geological methods of dating, we now know the timing and pattern of many extinctions in considerable detail. Globally, the chronological progression is extremely interesting. The sweep of extinctions of large mammals began gradually and inconspicuously in Africa over two million years ago, intensified in Europe beginning with the extinction of the Neanderthals 50,000 years ago, hit hard in Australia 40,000 to 50,000 years ago, and exploded in the Americas around 13,000 years ago. Apart from the Tertiary genus *Imagocnus* of Cuba, all extinctions of dwarf ground sloths in the Greater Antilles that can be dated are considered to be postglacial, postdating the extinctions of their relatives, the megalonychid ground sloths of the embracing American continent. Dwarf ground sloth extinctions in the Greater Antilles began around 5,000 years ago. In the last 3,000 years, extinctions swept thousands of taxa of flightless birds and many land snails from the islands of the remote Pacific, beginning in the west in Tonga, New Caledonia, and Fiji (those islands closer to the Asian mainland) and ending 1,500 years ago (or later, in some cases) in Hawaii, the Marquesas, and Rapanui (Easter Island). Extinctions struck Madagascar beginning less than 3,000 years ago. The extinction of moas in New Zealand took place only 500 years ago. These extinctions reflect the spread of our species.

There are far fewer deep-water islands in the remote Atlantic than in the remote Pacific, and very likely for that reason prehistoric voyagers did not manage to find and colonize the Azores, Ascension, Bermuda, Fernando de Noronha, and St. Helena. Extinctions in radiocarbon time generally occurred later on the remote Atlantic islands than on the deep-water islands in the remote Pacific, which had been discovered and in many cases settled prehistorically, beginning 3,000 years ago.

Historic extinctions include at least 200 species of vertebrates (Flannery and Schouten 2001), ranging in size from Steller's sea cow (10,000 kilograms, or 22,000 pounds) on islands in the Bering Strait to Lyall's Wren (flightless and mouse-sized) on Stephen's Island in New Zealand. Even before the loss of the moa, the introduction of Pacific rats apparently eliminated many ground-nesting birds and giant insects in New Zealand (Worthy and Holdaway 2002). Endemic species on oceanic islands remain at risk, as the recent fate of flightless rails and other birds subjected to predatory attack by introduced continental brown tree snakes (Boiga irregularis) indicates on the island of Guam (Wiles and others 2003).

These tail-end, historic extinctions have been a replay of the big event in miniature (Flannery and Schouten 2001; Quammen 1996). They are far outnumbered by earlier near-time extinctions, which, overall, eliminated large land vertebrates to a degree not seen since the end of the dinosaurs.

To me the core piece of evidence for human involvement is that when viewed globally, near-time extinctions took place episodically, in a pattern not correlating with climatic change or any known factor other than the spread of our species. Extinctions followed prehistoric human colonizations in a "deadly syncopation," to use the words of mammalogist Ross MacPhee. As explained further in "Radiocarbon Dating and Quaternary Extinctions" (p. 44), there is radiocarbon and other geochemical evidence that the earliest human arrivals on various landmasses were contemporaneous with the last days of the extinct species. Simply stated, as humans moved into different parts of the planet, many long-established huntable animals died out.

The archaeological record indicates that the earliest humans in America were the Clovis people (so named after their distinctive spear points, first found near the town of Clovis in eastern New Mexico). Around the time Clovis points and other prehistoric artifacts first appeared, two-thirds of the large animals of North America north of Mexico suffered explosive extinction. Mexico, Central, and South America lost even more, including commensals and parasites that disappeared with their hosts. Extinctions in other parts of the planet, too, are suspiciously close to the time of first human arrival. The scenario is familiar to us in the case of historic extinctions. We can document that many of these were caused by over-hunting, habitat destruction, or the introduction of other aliens. Nevertheless, many archaeologists are rooting for a pre-Clovis entry of humankind. In addition, some prefer to attribute the megafaunal extinctions to climate change. At times the dispute resembles the partisan passion of a hard-fought political campaign. Some archaeologists are baffled that humans can be held responsible for extinctions of large mammals, such as ground sloths and glyptodonts, only 13,000 years ago, with no evidence of even a single kill site. The problem is that the fossil record rarely discloses the cause of mortality, much less of extinction.

In near time, Africa and Asia suffered least. This does not mean that hominids would have had little to do with shaping the African and Asian faunas. However, interactions between evolving hominids and African wildlife, for instance, extend over at least two million years. The resulting coevolution would have meant that as human hunting skills advanced, so did the wariness and defenses of potential prey. Similarly, in Eurasia, the contact between horses and camels and Paleolithic people was gradual and thereby more favorable for both survival and domestication of the large herbivores than the sudden sweeping contact that took place when humans entered North America.

The evidence regarding the mammals' final demise is quite different from that regarding the dinosaurs'. As every first-grader knows, dinosaurs vanished 65 million years ago, smashed to pieces by a space rock. Not all geologists will agree, but I am impressed with the evidence that dinosaur extinction may have involved an asteroid or large meteor striking our planet (L. Alvarez and others 1980; W. Alvarez 1997; Powell 1998), an extraterrestrial accident that has been compared with the detonation of a 50-ton atomic bomb. Whatever happened also wiped out innumerable other terrestrial and marine species, some of them small in size, along with many vascular plants. In near time, however, there is no hint of such a catastrophe. A meteor crater in northern Arizona is about 50,000 years old, too old and much too small to account for extinctions of mammoths and ground sloths. There is no evidence of elements that are common extraterrestrially but less so on Earth, such as iridium, and no sign of a tsunami or other phenomena following the impact. Moreover, the near-time extinctions were highly selective, sparing aquatic species and plants. Finally, their unfolding over tens of thousands of years precludes a one-shot global catastrophe. Indeed, perhaps the one thing most specialists can agree upon is that the near-time extinctions had nothing to do with a space rock. In a sense, the absence of such an explanation is unfortunate, as one would find a highly receptive audience among astronomers and their public had mammoth extinction shared similarities with dinosaur extinction.

As noted, climate change is frequently invoked to explain the near-time extinction spike. Geological evidence and shifting percentages of fossil pollen indicate that the late Quaternary was indeed a time of severe and rapid climate change. So, however, were earlier stages of the Quaternary. Why, then, is there not a trail of extinctions of large mammals through the last two million years as the ice advanced and retreated, sea levels rose and fell, and plant communities moved farther north and farther south, higher and lower in elevation? If glacial climates forced the large animal extinctions, they should have struck when the susceptible animals encountered unfavorable climatic change going into the Quaternary. This is what paleontologist O. P. Hay claimed early in the twentieth century, before it could be shown that supposedly "early" Quaternary faunas were actually misdated (Martin 1967b). They were much younger. Extinctions of large Quaternary mammals in North America did not concentrate toward the beginning of the Quaternary ice age, or throughout the 1.8 million years involved, but piled up toward the end, within near time. Whatever was involved in forcing extinctions had to be something late in the Quaternary, not early or throughout the ice age. The possibility arose that the entire concept of climatic change as a driving force in prehistoric extinctions might be bankrupt. That would not go over with many vertebrate paleontologists, including one of my former dissertation committee members at the University of Michigan, Claude Hibbard.

Known to his friends as "Hibby," he prepared a surprise for me when I gave my first extinction talk to the Museum and Department of Zoology at the University of Michigan soon after earning my Ph.D. Hibby believed that the fossil giant tortoises of Kansas, where he excavated spring deposits, could not have survived freezing temperatures in winter because their living representatives, Galapagos tortoises, died when exposed to cold weather. After I finished my review of the large animal extinctions, having showed how on a world scale they corresponded with human arrivals, and called for the lights, Hibby whispered to the projectionist to hold the lights while he slipped in a slide of his own. Up on the screen went a cartoon of a glacial winter in Kansas, complete with a giant tortoise decked out in a scarf, earmuffs, down jacket, and snowshoes. The audience roared. Hibby won that round. Nevertheless, we cannot be certain that giant tortoises in North America did not have the physiological capacity to survive the coldest winters deep underground, like early homesteaders in the west who dug their cellars first so they could winter there.

The near-time extinction spike in the late Quaternary must have been caused by something outside the normal experience of mammals, and climatic change is as good a place to start the analysis as any. But the temperate animals that suffered extinctions had long been exposed to continental climates and presumably must have been accustomed to many changes. According to John Alroy (1999), there had been no more severe loss of large species since the evolution of large mammals tens of millions of years earlier. The arguments for and against climate change as the cause of the extinctions are discussed more fully in chapter 9.

The cause of prehistoric extinctions has been a topic of hot debate since the mid-1960s, debate that I helped ignite by linking the extinctions to the first appearance of humans—Clovis hunters, or First Americans—in the New World (Martin 1958a, 1967b). Some anthropologists find merit in the overkill theory, others discount or reject it because fossil evidence for human hunting in the crucial time frames is sparse at best. Some argue that even if human predation were involved, climatic change must be as well (Krech 1999).

Moreover, the proposal that near-time extinctions in some critical way involve people, our species, *Homo sapiens*, requires at least a modicum of cultural sensitivity. Certainly no one can pass judgment, from long after the fact, on the peoples who first discovered and inhabited new lands. Their achievements were truly remarkable. It is one thing to note synchronicity in the arrival of first pioneering prehistoric people in various corners of the planet and the concurrent extinction of many native animals; it is another to make a judgment. It would be absurd to assign blame to the progeny of Paleolithic Europeans or of the First Americans for the extinction of the Old World or New World mammoths, to Australian Aborigines for the end of the diprotodonts, or to the New Zealand Maoris for eliminating the moa. It is important to remember that the extinctions of near time occurred worldwide. To the extent that responsibility is assigned, it belongs to our species as a whole. This may be an even more disturbing thought for many.

Geologists travel into "deep time," which envelops a fossil record of hundreds of millions of years of organic evolution, including five mass extinctions. The rest of us may regard events of 13,000 years ago (the time since the American megafauna disappeared) as decidedly ancient. After all, a life span of one century is beyond the reach of all but a very few of us. Who can comprehend 130 centuries, over 10 times the age of Methuselah? In a sense we are like fruit flies, which live but a few weeks and cannot experience most seasonal changes, much less a year. We cannot know from experience the history of planet Earth. Most of it is destined to be as abstract to a layperson as the dimensions of the universe.

Why, then, is it important to understand what happened to the large mammals thousands of years ago? Why should we care that they are no more, and that our species might be the reason for this? Since they are extinct and thus absent from our zoos, who really misses the giant short-faced bears? Or the giant ground sloths? And why should even those particularly interested in the survival of species look to a distant, irretrievable past when the present and the future offer so many immediate concerns?

Such reductionism shuts off vital insights and golden opportunities. Extinctions in near time look suspiciously like the overture to the accelerating extinctions of recent years. For several reasons, extinctions in near time cast a very dark shadow on ecological systems worldwide and on efforts to protect those systems.

First, if we do not look back to the late Quaternary, we underestimate the rate of extinction during the presence of humans on the planet. This is no problem for normal or background extinctions, which sputter along like the decay of isotopes. Theoretically, background extinction is roughly in balance with the evolution of new taxa. But extinctions in near time far exceed background extinctions. Ignoring, for example, the disappearance around 13,000 years ago of the horses, mammoths, and mastodons that had been native to North America for tens of millions of years seriously affects any estimate of the rate of environmental degradation during our tenure on the planet. What would it mean to someone measuring the rate of theft from a department store if many of the large and expensive items—the overstuffed couches, recliners, dining room tables, and giant-screen TV sets—had been ripped off shortly *before* the inventory began? Surely this pilferage would have to be taken into account.

Second, the loss of these mammals before historic time crippled our reaction to recent extinctions and threats of extinction. The absence of mammoths, ground sloths, and others derailed a much more intense involvement with American wildlife than could be developed with the blighted survivors of near-time extinctions. Most of us are more susceptible to large, warm-blooded, furry, bright-eyed mammals than to reptiles, amphibians, insects, birds, or tiny mammals such as shrews. Perhaps it is because the large mammals seem to be the most like us. It is no coincidence that so many conservation organizations choose such animals as the symbols for their campaigns. Lewis and Clark should have found "great claw," just as Jefferson hoped.

It is entirely possible that even if the amazing late-Pleistocene megafauna had survived on our continent to the time of European arrival, we would have succumbed to the nineteenth-century delusion of limitless wilderness and hunted most of them down, the way we did bison in the 1880s. However, had even a few of the gentle giant ground sloths and their coevals survived to the twentieth century, a campaign to save them would surely have had more mass appeal than the campaigns to save the snail darter and the Delta smelt, ecologically important though these small fish are. The very existence of the great mammals might have helped us come to our senses. Impulses for conservation and preservation could have been awakened far earlier and more powerfully, and I believe we would be far ahead of our current efforts to attain global sustainability, heroic as they are. A crucial corollary is that greater awareness *now* of these long-extinct mammals may still energize conservation efforts. Knowing that we have lost far more than we had ever imagined makes it even more vital to preserve what we have left.

Third, ignorance of the late-Pleistocene extinctions warps our view of what "state of nature" we should be trying to conserve or restore. In North America, the modern extinction-pruned large-mammal fauna, those animals at "home on the range" since European settlement, are not a normal evolutionary assemblage. The fossil record thus suggests, for instance, that we reconsider the impact of wild equids in the New World. Because horses evolved here, flourished for tens of millions of years, and vanished around 13,000 years ago, their arrival with the Spanish in the 1500s was a restoration, not an alien invasion. In evaluating the ecological impacts of wild horses and burros, we need to be aware not just of their presence in the last half millennium, but of the coevolution of equids with the land for tens of millions of years before a relatively brief 10,000-year interruption.

More broadly, the life and times of the mammoths and other megafauna need to be understood before we can claim to know the true nature and potential of our planet. If Henry David Thoreau was not thinking of mastodons and giant ground sloths when he wrote, "I wish to know an entire heaven and an entire earth," he might as well have been. The last "entire earth" known to humans disappeared with the mammoths around 13,000 years ago. Thoreau wrote in his journal (March 23, 1856): "I cannot but feel as if I lived in a tamed and, as it were, emasculated country. . . . I should not like to think that some demigod had come before me and picked out some of the best of the stars." Demigods (the first European settlers) had driven cougar, lynx, wolf, moose, and deer from the woods of Concord. And earlier demigods, the First Americans, may have destroyed the mastodons and other large Pleistocene mammals.

If so, Thoreau's words reached deeper than he knew. Because of our ignorance of, or indifference to, prehistoric extinctions, we lack understanding of the natural riches of this planet, and that lack has narrowed our vision of how to plan for the future, especially in America.

After 40 years of investigating the late Quaternary, I am dismayed that the fate of the Quaternary megafauna and its meaning for the deep history of our country, our continent, and our planet have yet to capture the public's interest. Most people visiting our great museums breeze by the mammalian megafauna to ask after the dinosaurs, even bigger and vastly more popular icons. We know that the extinction of the dinosaurs was an awesome event. My purpose in writing this book is to encourage similar awareness of the extinction of the great mammals.

The following chapters outline the evidence, as I see it, for the overkill theory. Some of this evidence comes from my own expeditions into near time, and more of it from the discoveries of former students and colleagues. We shared adventures as we investigated the extraordinary animals that once lived on the planet and how they fared when people began to spread out of Africa and Eurasia. For me, these adventures began at the Desert Laboratory of the University of Arizona—a unique source of new data and a sanctuary against too much exposure for those of us testing radically new ideas about the wild.

THREE

GROUND SLOTH DUNG
AND PACKRAT MIDDENS

The greatest impediment to scientific innovation is usually
a conceptual lock, not a factual lack.

Stephen Jay Gould, *Wonderful Life:*
The Burgess Shale and the Nature of History

In 1956, after two years on fellowships, it was time to look for a permanent teaching and research job. My top priority was finding a position that would give me an opportunity to recover a fossil pollen record
next to radiocarbon-dated remains of extinct animals of the ice age. That
combination of data should shed light on the causes of the extinctions
by showing whether or not climatic changes correlated with them.

Yale conservation ecologist Paul Sears and his students had recently
discovered important climatic changes in the pollen record from a long
core recovered from lake sediments beneath Mexico City. In addition, in
a long core taken in the San Agustin Plains of New Mexico, Sears and
Katherine Clisby had found pollen evidence of a glacial-age spruce forest well below the elevations now occupied by spruce in the southern
Rockies. Unlike pollen records inside the glacial ice margin of New England and the Great Lakes region, these cores yielded continuous records
well beyond the latitudes covered by Quaternary glaciers.

From these and other findings I concluded that the southwestern
United States was a promising place to look for fossil pollen deposits dis-
closing plant and animal range changes forced by climatic shifts in the
ice age. The floodplains of Cochise County, Arizona, for example, offered

abundant alluvium (sediment laid down by moving water) that invited coring and analysis. Some arroyos had yielded bones of mammoths and associated artifacts. In addition, the buried muds of the dry playas (basins) left by Quaternary lakes were rich in fossil pollen coinciding in time with the Quaternary megafauna. The 1,000-foot core Clisby and Sears took from the San Agustin Plains provided one of the first tests in the arid West of the wrongheaded interpretation that lower latitudes escaped the climatic influence of the last ice age. They discovered appreciable percentages of glacial-age spruce pollen in their core, collected in an environment dominated by juniper and pinyon, with ponderosa pine growing only on the most favorable soils. No longer was there any doubt that when glaciers advanced and retreated, plant and animal communities, spruce included, did as well. Radiocarbon dates demonstrated that all these changes were synchronized.

The arid Southwest also offered another crucial advantage, from my perspective: excellent preservation of datable fossils, especially in caves. Except in frozen ground, fossils buried in open sites are leeched of their organic content and are rarely datable by isotopic measurement. On the other hand, investigators have been aided by the remarkable preservation of bones in dry caves in arid regions, as well as in unglaciated boreal or subarctic regions with permafrost. In certain caves in the Grand Canyon, extinct bird and mammal bones, retaining organic residues such as collagen and in some cases with dried tissues still attached, were to prove ideal for radiocarbon dating. Also ideal are keratinous tissues, such as beaks, claws, hooves and horn sheaths, which decay rapidly in temperate and tropical environments but not in dry caves or rock shelters in arid regions. Skin, hair, cartilage, and even dung balls of long-extinct species can be remarkably well preserved. Recently such deposits have proved rich in well-preserved and identifiable DNA of the animals once inhabiting the caves.

All in all, the Southwest seemed like the very best place for me. Unfortunately, I could find no university positions open in Arizona or New Mexico, where I wanted to live. Then a friendly Arizona zoology student pointed out that universities might give research posts to faculty who came with their own grants, such as those from the newly established National Science Foundation (NSF). This proved to be excellent advice. The University of Arizona's preeminent anthropologist, Professor Emil Haury, director of the Arizona State Museum, put me in touch with Terah (Ted) Smiley, director of the university's new geochronology program. Although we had yet to meet, Ted and I collaborated by mail on a grant proposal involving pollen analysis of archaeological sites with extinct animal remains. I suspect that Ed Deevey and/or Paul Sears reviewed our NSF proposal. In any case, we won an award that paid my salary and that of two research assistants for two years. In the fall of 1957 my wife Marian and I and our three small boys, Andy, Neil, and Tom, left our families on the East Coast and headed for Tucson.

I had not heard of the university's historic and world-famous Desert Laboratory on Tumamoc Hill. Built in 1903 by the Carnegie Institution of Washington, the laboratory was initially dedicated to desert botanical research (McGinnies 1981; Bowers 1988), such as studies on whether saguaros were disappearing from the vicinity of Tucson (they were not). Pioneer desert botanists, such as Forrest Shreve, who helped to found the Ecological Society of America, spent much of their careers here (Bowers 1988). In the 1930s, the Carnegie Institution of Washington abandoned several of its earlier programs in favor of more advanced "experimental" sciences. Under Ted Smiley's leadership, however, the Geochronology Laboratories carried on the tradition of research on arid regions, incorporating fossil pollen analysis, vertebrate paleontology, geomorphology, organic and inorganic geochemistry, and paleomagnetism. On Tumamoc Hill, Ted found space for offices, a conference room with library, a vertebrate fossil preparation lab, a hood and bench space for pollen extraction, and the magnetometer of a geophysicist.

The university soon purchased the Desert Laboratory and its grounds plus leases, a total of 869 acres of the Sonoran Desert. The grounds harbor 350 species of vascular plants, 300 of them native (Bowers and Turner 1985), as well as numerous mammals, birds, and reptiles.* (Some of this area would make excellent camel habitat.) From the lab's roost on a bench at 2,400 feet, one can look out over Tucson into three of Arizona's "sky island" mountains, each close to or exceeding 9,000 feet in elevation and each supporting rich vegetation gradients. Thanks to seminal work by the Carnegie botanists, the lab features a rich legacy of long-term studies. The pleasure of working in such a place would be—well, I hoped that University of Arizona administrators would not subtract the considerable value added by a desk and lab space on Tumamoc Hill from my paycheck.

Not wasting time on formalities, on my first day on "the Hill" Ted

*For a state-of-the-art treatment of shrubs of the Sonoran Desert and the geomorphology near Tucson, see McAuliff 1999.

waved me to a chair and handed over a large brown paper bag labeled "Rampart Cave." Inside I found many smaller bags of samples collected by a graduate student in anthropology, Dick Shutler Jr. Setting aside his pipe, Ted reached into one of the bags, handed me a fibrous brown object the size of a baseball, and asked if I had seen anything like it before.

Taken aback, I received the segmented bolus of tightly packed dry plant remains somewhat gingerly. Although it was too large and misshapen to be a "road apple," the dung ball of a horse or a mule, that was the closest I could come. Ted's eyes twinkled. He explained that it was ground sloth dung or coprolite (fossilized dung). (In polite company in those days we did not say "shit.") From a cave in the lower end of the Grand Canyon, Dick had collected the dung samples at 6 inch intervals from top to bottom of a 5-foot stratified section. Years later we could see where he had removed samples, one above the next, from the trench through the ground sloth dung. Shasta ground sloth *(Nothrotheriops shastensis)* bones were associated with it. (A virtually complete skeleton of a Shasta ground sloth, with a dung ball in the paunch area, had been discovered in the 1920s at the bottom of a fumarole near Aden Crater, New Mexico; this confirmed the identity of the Rampart Cave dung.) Although compact and hard, the specimen was not mineralized or calcified. I ventured a cautious sniff. Though it was almost odorless, it looked fresh, fresh enough to have come not from an extinct animal many thousands of years old, but from one that might be still alive!

Ted could not have had better bait to capture my interest. By radiocarbon assay on the dung we could determine when the ground sloths lived in Rampart Cave and, along with results obtained elsewhere, estimate the time of their extinction. Because the samples were stratified, they would differ in age, older at the bottom, younger on top. We could determine the rate of deposition.

In addition, I was certain the dung would contain fossil pollen. Finnish geologist Martti Salmi had successfully extracted pollen from a single dung ball of a South American ground sloth (Salmi 1955). Digestion does not destroy the morphologically distinctive walls of pollen grains or of fungal spores, which resist hydrofluoric acid, capable of dissolving silts, clays, and even quartz. With any luck we could do the same and recover the first profile from a stratified dung deposit of an extinct late-Quaternary herbivore. Through the fossil pollen and other plant remains, we could detect changes in vegetation, climate, and the diet of the ground sloths. Although stable isotopes may be of some help, the diets of extinct animals are impossible to determine in any detail from fossil bones. Unmineralized dung deposits are ideal but so unusual that most paleontologists never see them. This is one of the factors that made Rampart Cave such a uniquely valuable site.

Finally, we could use trace element geochemistry to look for essential mineral nutrients in the dung. Salmi had proposed that ground sloth extinction was driven by scarcity of essential trace elements, such as copper and cobalt. We could test that.

Ted said I could share his office space in the library, and I proceeded to move in. To varying degrees academics lead nomadic lives. For seven years my family and I had been on the move. I did not know it on that mellow, mild November day over 45 years ago, but we had come home. I soon began to teach classes at the Desert Lab, which was just far enough from the main campus to discourage the less-than-dedicated students. In general I found motivation a far better predictor of career and personal success than grades on a transcript. Those willing to travel some 20 minutes by auto (longer by bike or bus) from the main campus to take a course or do research were sure to be worthy.

Dick Shutler let me join his Rampart Cave ground sloth project, and our geochemist colleague Paul Damon suggested that we add his student Bruno Sabels to the team. Dick soon left to receive state-of-the-art training at a laboratory affiliated with Columbia University, one of the best radiocarbon laboratories in operation at the time. He took splits of his sloth dung samples with him to have them dated. As discussed further in chapter 4, his radiocarbon dates indicated that the youngest dung was roughly 10,000 years old.

Meanwhile, I extracted and counted pollen from the samples. The pollen profile reflected the diet of the Shasta ground sloth. There was an abundance of globe mallow, *Sphaeralcea* sp., which is in the hollyhock family. Globe mallows are not wind pollinated, so their pollen would not have blown into the cave; the sloth must have eaten flowering plants. In favorable years *Sphaeralcea* flowers heavily in spring, presumably the time the ground sloths occupied the cave.

That spring I noticed a native globe mallow, with its bright orange-red flowers, growing around the Desert Laboratory. Emulating a ground sloth, I chewed some of the tops. The leaves were rather bland and slightly slimy, tasting like a young hollyhock, not bitter or resinous like many desert shrubs, such as creosote bush. With a hand lens I could detect indigestible star-shaped hairs known as trichomes on the foliage of the mallows. An abundance of the same type of hairs as well as pollen in the sloth dung supported the conclusion that the animals had been eating globe mallows.

Not all interpretations are as straightforward. In some extractions from the sloth dung, juniper exceeded 90 percent of the total pollen. But unlike the case of the mallows, this did not necessarily mean that the ground sloth ate juniper. Unlike globe mallow, juniper is wind pollinated. In a wet year juniper trees are loaded with ripe anthers discharging clouds of pollen, dusting all plants in the vicinity. Under those circumstances, any animals eating plants growing anywhere near juniper would unavoidably ingest an abundance of its pollen.

The presence of so much juniper pollen indicated the presence of junipers themselves and a change in climate over the last 30,000 years (Martin, Sabels, and Shutler 1961). From Carnegie botanical publications (Laudermilk and Munz 1934, 1938), verified by our own observations in later years, our field team established that junipers do not grow near Rampart Cave now, but the pollen record certainly suggested that they had in the past. (The same is true regarding Gypsum Cave, Nevada, where we later studied another dung deposit, as discussed in chapter 4.) Furthermore, given the season when juniper trees release their pollen, I concluded that the Shasta ground sloths visited Rampart Cave during winter or early spring.

Using mass spectrometry, Bruno Sabels found that the ground sloths at Rampart Cave had ample copper and other trace minerals in their diet. Salmi's suggestion that scarcity of trace elements caused the ground sloth extinctions had seemed questionable. According to Simpson's checklist, in the late Quaternary a dozen genera of ground sloths vanished from North and South America, along with or followed by dwarf ground sloths from the West Indies (see McKenna and Bell 1997 and White and Mac-Phee 2001 for updates; see Lyons, Smith, and Brown 2004 for a species list). It seemed unlikely that the entire New World had somehow run out of essential trace elements, even if southern Chile had done so. The soils in many parts of the Americas were not lacking in essential minerals. Copper, for example, is so abundant that Mexico, Arizona, New Mexico, Utah, and Montana, as well as the Andes in South America, host giant open-pit copper mines. Sabels's geochemical results proved that whether or nor Salmi's explanation for ground sloth extinction might apply in southern Chile, in Arizona it lacked support.

Not all of Dick Shutler's samples were of dung. The contents of one of his collections included fecal pellets of packrats and plant material brought into the cave by packrats. This might have revealed a new source of data for assessing climate and range changes, if we had only recognized its significance. In the early 1960s, ecologists Phil Wells and Clive

Jorgensen would report a surprisingly rich and previously unknown or underappreciated source of fossils: the fossilized middens of packrats *(Neotoma)*, also known as wood rats (Wells and Jorgensen 1964). We had failed to recognize how abundant packrat middens can be in caves and rock shelters in most of the arid West.

Twenty-one species of packrats range from western Canada south to Nicaragua, with seven species in the Southwest of the United States (Vaughn 1990). True to their name, packrats throughout the ages have collected items of interest to them, often, but not always, items they could use for nest construction. Active packrat nests in open sites, often under prickly pear *(Opuntia)* or other cacti, are bushel-basket-sized piles of sticks, leaves, cactus pads, stones, and stored plant food accessible by hidden runways. Such middens are familiar to many hikers and naturalists, especially in arid America. In open sites, inactive middens do not last very long. However, middens in caves or rock shelters may harbor abandoned deposits that, saturated with packrat urine, can harden into resinous lumps as hard and durable as adobe (see plate 1). These fossil middens can last for thousands to tens of thousands of years, as long as they remain dry and are sheltered from precipitation, runoff, and incursions of termites (see various chapters in Betancourt, Van Devender, and Martin 1990). Similar deposits left by small mammals are reported in arid regions of South Africa, Australia, South America, the Near East, and Mongolia.

Each midden is a veritable time capsule, containing a sampling of whatever materials were available to packrats at the time the midden was active. Middens in caves are sometimes as small as a brick, sometimes a bit larger than a cement block, and contain leaves, stems, flowers, dry fruits, seeds, and pollen, along with an abundance of sticks, cactus pads or aureoles, packrat fecal pellets, and occasional bones or droppings of other animals, both living and extinct. Because plant fossils in middens are not mineralized, detailed anatomical analyses of leaves and other tissues can be made and remnant DNA identified.

The ancient rat middens proved to be a paleoecologist's bonanza. The contents of a single midden represent a very brief interval in near time. With accelerators one can directly date the remains of any species of woody plant found in a midden. Multiple radiocarbon dates can test the assumption of a midden's chronological integrity. Middens may be stratified, incorporating levels of different ages as determined by radiocarbon dating.

Rich in seeds and other identifiable macrofossils of many arid land trees and shrubs, fossil middens reveal the species composition of pre-

Plate 1. The author at a glacial-age packrat midden with extralocal juniper twigs, Montezuma Head, Organ Pipe Cactus National Monument, January 1976. Photo by Hal Coss.

historic plant communities, including some once inhabited by the extinct fauna of the late Quaternary. Added to fossil pollen records from desert playas, springs, spring mounds, and mountain lakes, the midden records permit the construction of maps showing the near-time ranges of many woody plant species and even some herbs (Betancourt, Van Devender, and Martin 1990; Brown and Lomolino 1998; Thompson 1990). Best of all, the middens were a new source of glacial-age plant fossils in dry climates, where stratified lake deposits were less abundant and often truncated. Though the combination of middens and fossil pollen data is especially informative, middens proved to be far more than a check on other approaches. They opened doors to new discoveries of biotic and climatic changes in near time.

The environments known to the extinct megafauna and to prehistoric people of the arid West could now be sampled in far more detail than had been possible previously. Stratified middens or those of different ages show changes in local ecology over time. They therefore provided a powerful new method of characterizing near-time environmental change and—when animal fossils were available—its relationship to faunal change, including late-Quaternary extinction. It was magical. We found ourselves time traveling.

Paleoecologists were by no means the first to appreciate fossil middens. A century and a half ago, about 60 miles northwest of Las Vegas, miners traveling west entered a canyon south of a dry playa known as Papoose Lake. They were out of rations and searching for food. Then, as one of them wrote: "Part way up we came to a high cliff and in its face were niches or cavities . . . in some of them, we found balls of a glistening substance looking like pieces of variegated candy stuck together. . . . It was evidently food of some sort, and we found it sweet but sickish, and those who were hungry . . . making a good meal of it, were a little troubled with nausea afterwards" (Manly 1987). A little nausea seems a small price to pay for "a good meal" of what looked like candy, "glistening" with ancient rat urine. Intrepid as they are, I do not know of any recent investigators who claim to have emulated the hungry miners. By coincidence, Wells and Jorgensen made their discovery within roughly 30 miles of Papoose Lake, in a cave on Aysees Peak, a desert mountain in southern Nevada near the Atomic Energy Commission's nuclear test site.

Our own near miss in recognizing the significance of fossil middens was much less dramatic. Dick Shutler's sample from 36 inches at Rampart Cave contained aureoles of hedgehog cactus (*Echinocereus*) and a juniper (*Juniperus*) twig, both part of a packrat midden. Since juniper did not grow near the site, the stratigraphic location of the twig supported fossil pollen evidence of a major change in vegetation, and presumably climate, during the last continental glaciation. However, we did not register that the twig was in a fossil midden, so we missed its larger implications. I wonder how many other ecologists, archaeologists, and paleontologists did the same in their field work.

It turned out that second only to the sloth dung itself, most of the organic fill of Rampart Cave represented middens or scattered plant material, both deposits of prehistoric packrats. Packrats living in the cave would have collected all of their food and nesting material within roughly 50 to 100 feet of its mouth. The rats left sizeable middens beneath or above rock ledges; within the 36-inch level, apparently during a lengthy absence of the ground sloths; and on the surface of the dung deposit, evidently after the extinction of the ground sloth. A significant fraction of the middens contained juniper, in some cases mixed with fecal pellets and horn sheaths of extinct mountain goats and bones of these and other extinct mammals (Phillips 1984).

In the fossil middens from Aysees Peak, Wells and Jorgensen found abundant juniper twigs and leaves (needles). Juniper often occurs in desert mountains, but despite a thorough search, Wells and Jorgensen could find none growing on Aysees Peak. They sent the juniper remains to Libby's radiocarbon dating laboratory at UCLA and reported, "The result, 9320 ± 300 B.P., supported expectations. Ten other middens dated between [40,000 and 7,800 BP] indicated that juniper's lower limit in the region was then at least 600 m [2,000 feet] below where it is now" (Wells and Jorgensen 1964, 11/3). They had nailed down the first of what would become an avalanche of new fossil discoveries. Midden analysis soon revealed glacial-age distributional shifts in a great variety of plants, including trees, shrubs, succulents, grasses, and herbs. Thanks to packrats we could determine, at least in general, the nature of the environment inhabited by the extinct animals of near time.

Full appreciation of these findings requires an understanding of vegetation gradients, formerly known as "life zones." A life zone is a geographic area in which a distinctive group of plants and animals is typically found. Of these, the trees and shrubs are generally the easiest to spot and to identify in the fossil middens. Each life zone typically occurs in a particular latitudinal range, where the appropriate climate prevails. The zones vary with altitude. In time the field ecologist will appreciate that what one sees is a continuous gradient rather than discrete zones. In the Northern Hemisphere, a higher elevation farther south may be as cool as, and thus support much the same vegetation as, a lower elevation farther north. In fact, it is often much easier to see the different zones by going 3,000 feet up or down a mountain than by driving several hundred miles north or south at the same elevation.

Differences in disturbance, substrate (bedrock and soil type), and aspect (solar exposure) introduce important complications in what actually grows where. Together with climate, elevation, and topography, we encounter a fistful of variables. (No wonder so many ecologists turn to modeling—it is the only way they can hope to keep up with nature's complexity.) These effects can be dramatic. For example, in the Northern Hemisphere, slopes facing north are relatively shady and thus cooler and less dry than south-facing slopes, which are fully exposed to the sun. Accordingly, a given plant will grow at different elevations on the different exposures. On the north-northeast side of a desert mountain, the lower limit of most plants is roughly 2,000 feet below their lower limit on the south-southwest side. That is, moving at a constant level from the south-southwest to the north-northeast side is ecologically equivalent to going up 2,000 feet in elevation.

The life zones form a magnificent gradient in which each species has its own unique habitat or niche. Different species may commonly be associated, but no two have identical distributions. For example, Engelmann spruce and alpine fir are often but not always associated in a spruce-fir community. Descending in elevation, these overlap with or give way to white fir, aspen, Douglas fir, limber pine, and ponderosa pine. Below, the ponderosas give way to pinyon-juniper woodland, typically with more pinyon at the upper elevations, while lower still, juniper may be accompanied not by pinyon but by single-leaf ash *(Fraxinus anomala)*.

Superb examples of elevational gradients can be found in the Grand Canyon, which rises from 1,500 to 3,000 feet at the bottom to 9,000 feet on the Kaibab Plateau and 12,500 feet on the once-glaciated San Francisco Peaks outside Flagstaff. From Phantom Ranch on the Colorado River to the San Francisco Peaks, an elevational range of over 10,000 feet compressed into a distance of just under 80 miles, ecologist C. Hart Merriam in the late 1800s recognized the biological equivalent of a journey from the Mexican lowlands to Hudson's Bay in central Canada (Houk 1996).

By driving from Lees Ferry at 3,000 feet to the top of the Kaibab Plateau at 9,000 feet, one can see all but one of the life zones Merriam identified here.* At the bottom is the Lower Sonoran Zone, a desertscrub of blackbrush, shadscale, brickel-bush, and Mormon tea *(Ephedra)*, with occasional sagebrush and other Great Basin shrubs and some Mojave Desert cacti. Going west, one climbs into open juniper woodland, then into denser pinyon-juniper woodland (Merriam's Upper Sonoran Zone). Jacob Lake, at 8,000 feet, lies in the Transition Zone, represented here by a magnificent ponderosa pine forest with patches of aspen and Gambel oak. Proceeding southward to the North Rim and the highest and coldest parts of the Kaibab Plateau, one enters the Canadian Zone, where ponderosa pine yields to Douglas fir mixed with aspen. Finally, a forest of spruce and fir grows above 8,500 feet, in the Hudsonian Zone. The uppermost zone in Merriam's system, the Arctic-Alpine, is restricted to elevations above 11,500 feet, which in Arizona occur only at the top of the San Francisco Peaks. Here hikers find endemic ground-hugging perennial herbs in meadows above the tree line.

Nearby, the overall gradient remains quite similar, but the specific associations vary somewhat. For example, within the Grand Canyon, hikers from the South Rim to Phantom Ranch descend through pinyon, juniper,

---

*According to the U.S. Board of Geographic Names, "Lees" has no apostrophe (letter from R. C. Euler, May 2001).

and banana yucca communities, but just below the rim, shady north-facing notches shelter occasional Douglas firs. Below, on the treeless Tonto Platform, they find a wide bench distinctively darkened by blackbrush—but if one knows where to look, I am told, occasional ponderosa pines can be found ensconced in shady north-facing breaks. On the slopes above the Colorado River and below the Great Unconformity, an incomprehensible time gap of 1.2 billion years, grow low shrubs of the sunflower family, such as aster, desert broom, and brickel-bush. Finally, riparian mesquites and catclaw acacia *(Acacia greggii)* grow next to the river, though catclaw may grow in moist sands up to 1,000 feet above it.

Similar gradients occur throughout Arizona and the West as plants respond to regional changes in climate and moisture resulting from differences in elevation, latitude, or longitude. The existence of such gradients over time is one thing that makes it possible to assess climate change. In cooler periods accompanying glaciation, Northern Hemisphere plants move southward in latitude, lower in elevation, or both to maintain a suitable environment. Similarly, in warmer periods, plants move northward or higher. Thus, for example, fossil evidence that a particular species formerly grew at a lower elevation than it does now suggests that the climate has warmed since that time. (Obviously, plants cannot literally pick up their roots and move; it is a question of where their dispersed seeds flourish.) Fossil pollen had been providing evidence of such climatic shifts; packrat middens were to provide still more, with less uncertainty about what plant species were involved.

Once Wells and Jorgensen had opened the door, many investigators in the Southwest began focusing on middens. Our team was no exception. With a modest amount of funding from the National Geographic Society, the NSF, and other governmental agencies, an interdisciplinary "Corps of Discovery" dedicated to finding, analyzing, and interpreting these middens took shape at the Desert Lab.

Fossil middens are not always easy to find or identify, even if one knows how to look for them. Talented midden prospectors, however, have turned up at some unexpected times. One discovery particularly memorable to me occurred in early June 1969, on an Arizona Academy of Science research trip down the Colorado River through the Grand Canyon.

On our fourth night we camped at River Mile 108.5, just above Shinumo Rapids. The trip's organizer, historian Marty Link, had scheduled "campfire seminars," and that evening it was my turn. I held forth on secrets of the past, including the treasures to be found in ancient packrat middens. How wonderful it would be, I added, to find a series of middens along the biotic gradient from the bottom to the top of the Grand Canyon. They might reveal what happened to plants—and thus potentially to animals—at different elevations as the climate changed during and after the last ice age. I invited my listeners to search for middens, noting that junipers, for example, no longer grew along this reach of the river, but that just possibly, there might be middens nearby that harbored juniper twigs from the last cold interval at the end of the Pleistocene.

Before sunup the next morning, before I could stir from my sleeping bag, I found three teenagers in my face. Since first light they had been crawling over boulders and shelving rocks, looking beneath ledges and into crannies for old middens. Now they had rushed back for breakfast and a chance to display their find. They thrust fist-sized chunks of some hardened lumps of plant material under my nose. The chunks smelled resinous and slightly fetid, the distinctive odor of old packrat middens. I thought I saw the fecal pellets of packrats. "See those twigs?" one of the boys demanded. "Don't they look just like juniper?" I fished out my hand lens, peered closely at a few dusty brown stems, and had to agree that they did.

After breakfast a small group of us followed the eager young prospectors of Pleistocene secrets to their claim. Not far from camp and within a few feet of a pin marking an archaeological site, the boys showed us dry plant material wedged in the rocks beneath a small overhang. On some twigs we could see the tiny but diagnostic awl-shaped needles of juniper. No junipers could be spotted growing in the vicinity, certainly none within packrat foraging range. We were looking at a late-Pleistocene displacement. It had taken the self-appointed field team less than an hour to come up with pay dirt. First-time prospectors are rarely so successful. This was my introduction to Jim (James I.) Mead, a high school student from Tucson. Son of Albert Mead, former head of the Zoology Department at the University of Arizona, Jim had grown up roaming the mountains of southern Arizona with his buddies, searching for land snails, his father's specialty. Jim is now a professor of geology at Northern Arizona University in Flagstaff.

The Shinumo discovery (Van Devender and Mead 1976) portended a flood of new records of plant displacements in the Grand Canyon at all elevations, some from collecting sites so difficult to access that the field team, Ken Cole and Geoff Spaulding, had their drinking water flown in by helicopter.

Not all who tried their hand at midden prospecting decided to join the club. On another outing, one very promising student who searched for juniper in a large rock shelter surrounded by teddy bear cholla (whose fiercely barbed joints are especially favored by packrats for protecting their nests) emerged terribly covered with spines. Although he had found a few scraps of juniper, previously unknown in the Tucson Mountains, he reverted to his first love, the excavation of Pliocene gomphothere bones in southeastern Arizona, safely removed from the unforgiving cholla spines in certain rat middens.

This was an exciting time. Investigators including Julio Betancourt, Ken Cole, Pat Fall, Jim King, Cynthia Lindquist, Jim Mead, Art Phillips, Geoff Spaulding, Bob Thompson, Tom Van Devender, Phil Wells, and Jeff Zauderer turned up rich midden records from various parts of Arizona, New Mexico, and Texas, or in the case of Fall and Lindquist, in hyrax middens from Jordan. Radiocarbon dating of the middens enabled us to generate Quaternary vegetation maps (Betancourt, Van Devender, and Martin 1990) superior to those based mainly on fossil pollen records and biogeography (Martin and Mehringer 1965). As Wells and Jorgensen had anticipated, however, the middens strengthened the findings of those using fossil pollen for paleoclimatic investigations. At the University of Arizona, professors Vera Markgraf and Owen K. Davis combined both approaches.

To my knowledge, Pete Mehringer (who, like Jim Mead, had spent many hours of his youth exploring the desert) and Wes Ferguson, a faculty member at the Tree Ring Laboratory of the University of Arizona, preceded others in making a fundamental find. At an elevation of 6,000 to 7,000 feet in the Clark Mountains of California, near Las Vegas, they found, in what is now pinyon-juniper woodland, fossil middens that contained the preserved remains of bristlecone pine, limber pine, and white fir *(Abies concolor)* (Mehringer and Ferguson 1969). In Nevada, neither bristlecone nor limber pine now grows as far south or as low in elevation as the Clark Mountains. Radiocarbon dates verified that both occupied the mountains in the cooler climates of the late Quaternary; the white fir, too, had descended in elevation in glacial times. The fossil pollen record of spruce in cores from the Willcox Playa had given Pete and me indications of a similar descent, but the middens furnished the first macrofossils, evidence more secure than fossil pollen, which is vulnerable to long-distance transport, that in the last glacial age not only juniper but also other montane trees in the arid West grew at lower elevations than they occupy at present.

Like fossil pollen from sediment cores, fossil packrat middens eventually showed that during the last glacial episode, most woody plants throughout the West descended by at least 2,000 to 3,000 feet, significantly below the elevations that they occupy now. In the process they might migrate south by hundreds of miles. Any fossil deposit that was over 8,000 radiocarbon years old could be expected to contain displaced species. In addition to juniper and bristlecone pine, examples included spruce in West Texas (Van Devender and others 1977) and in the Grand Canyon (Cole 1990; Coats 1997); and fir, Douglas fir, limber pine, and sagebrush (various chapters in Betancourt, Van Devender, and Martin 1990). Fossil pollen from sinkhole lakes revealed that alpine sedges and grasses had replaced boreal trees, such as spruce and fir, on the Kaibab Plateau (Weng and Jackson 1999). In southern Arizona's Organ Pipe Cactus National Monument, whose low mountains now support a Sonoran thornscrub community of organ pipe cactus, saguaro, and foothills paloverde *(Cercidium microphyllum),* Tom Van Devender found middens over 8,000 years old yielding Mojave desertscrub species: sagebrush, Ajo Mountain scrub oak *(Quercus ajoensis),* Joshua tree *(Yucca brevifolia),* and on Montezuma Head at 1,100 feet, one-needle pinyon *(Pinus monophylla)* and juniper (Van Devender 1990). This is a community much more like that found in Joshua Tree National Park, which is to the northwest, in California, and higher in elevation.

Investigators have interpreted these range changes in Arizona and adjacent states as the result of cooler climate and/or increased precipitation. Later, in the warming postglacial climate, spruce forest replaced arctic alpine tundra above 8,500 feet, while at lower elevations pinyon-juniper replaced spruce–mixed conifer woodland and Mojave desertscrub replaced pinyon-juniper–single-leaf ash woodland.

Just to make the record more interesting, various plant species did not simply descend in lockstep in glacial times. The ranges of some plants expanded or shrank beyond what one might have expected, or switched direction. For example, shadscale, a common Great Basin shrub of arid climate, presently grows near Lees Ferry on the eastern edge of Grand Canyon National Park, and, sporadically, in the western end of the park near Rampart Cave. Fossil middens reveal that in the late-glacial climate, shadscale expanded its range in many directions, even upward (Spaulding 1990), seemingly against the tide of species coming down. No other desert shrubs are known to do this. A change in soil pH as well as climate may have been involved (Martin 1999).

Interestingly, Pete and Wes found no indications that ponderosa pine *(Pinus ponderosa),* the dominant species of Merriam's Transition Zone, had occupied the Clark Mountains with the other montane conifers. Ponderosa is widespread today, ranging from the Sierra Madre Occidental of northern Mexico through the Southwest and the lower elevations of the Rocky Mountains into western Canada. In the last late or full glacial, it might well be expected to have occupied lower elevations in many parts of the West. Instead, the middens suggest that it was absent from much of its modern range. Fossils of ponderosa pines more than 10,000 years old have yet to turn up in the midden record in the Grand Canyon (Cole 1990) or in glacial-age middens outside of Arizona. In the United States, ice age ponderosa appears to be limited to southern Arizona and New Mexico. (I am well aware that in making a sweeping statement about what has not been found, I am tempting the fates to deliver contrary examples. So be it.) Julio Betancourt, Tom Van Devender, and I are keenly aware of the "conceptual lock" rather than "factual lack" against our historical revision, which finds cultural, not climatic, history of critical importance in this case (Betancourt, Van Devender, and Martin 1990, 2).

Ponderosa continues to expand in the region, as shown by repeat photos by Ray Turner of the U.S. Geological Survey and others of originals taken over the last 100 years. The postglacial spread of ponderosa is so extraordinary that it challenges us to consider forcing functions beyond climatic change. That is, did something happen to favor ponderosa after the last ice age, something in addition to the climatic warming that led most forest and woodland species to ascend in elevation? A major change in fire history, including season and intensity of firing, is one possibility. Ponderosas are fire-adapted. And with the arrival of people in the New World around the end of the last ice age, a change in wildfire frequency could be expected. Julio Betancourt (personal communication, December 2001) suggests that ponderosa pine benefited by fires set by Native Americans, artificial ignitions of relatively light intensity, set well in advance of the normal summer lightning strikes and ignitions. By removing excess fuel in advance of the season when firestorms are likely to develop, cool fires could have favored the ponderosas.

The fossil record of Colorado pinyon *(Pinus edulis)*, too, is unusual. Colorado pinyon is presently widespread in the Grand Canyon and Colorado Plateau at elevations just below those of ponderosa pine. In the late Quaternary, it did not simply descend a few thousand feet like spruce, Douglas fir, and Utah juniper. Instead, it almost vanished from Grand Canyon National Park (Cole 1990). Then, along with ponderosa pine, it expanded in range in the postglacial, perhaps as a result of human ignitions. In Chaco Canyon National Historical Park, the record indicated prehistoric human as well as climatic impacts on pinyon distribution.

The midden harvest was rich indeed. A number of species of plants appeared in the fossil record for the first time, greatly increasing our knowledge of their temporal distribution. And the dynamic changes implied by the past distribution of desert trees and shrubs laid to rest any thought that the arid Southwest might have escaped the climatic changes of the late Quaternary. A century ago, many botanists believed that the southwestern deserts were too low in latitude and elevation to have undergone significant climatic change in step with higher-latitude glaciations. However, both fossil pollen and macrofossils of the common plants that attracted foraging packrats showed that the dramatic shifts in eastern plant communities uncovered by Ed Deevey's pollen lab and many others since the 1950s had their equivalents in the West. The information needed to understand global climate change in the arid West in radiocarbon time comes to us courtesy of long-dead packrats.

As the midden research proceeded over the years, I kept one eye open for any unprecedented change in climate or environment that might help to explain the megafaunal extinctions. The extinct animals certainly lived during a time of dramatic vegetation change in the last cold stage, beginning 23,000 years ago and ending 8,000 to 10,000 years ago. However, nothing that I could detect in the new paleoecological and geochemical data from either the West or the Andes, under study by Julio Betancourt, Jay Quade, and their students, suggested a unique climatic crisis that would account for a unique extinction episode.

In contrast, midden analysis did support an important argument in favor of the overkill theory. As the list of tree and shrub species known from middens accumulated, it became increasingly apparent that plants, unlike large terrestrial animals, had not experienced a wave of extinctions in the late Quaternary. The near-time fossil record of plants in the arid West is similar to that of beetles (Coope 1995). Those records show that both beetles and vascular plants are sensitive to climatic change, but that neither suffered appreciable near-time extinction. Presumably, then, neither group was vulnerable to whatever wiped out the large mammals. This is in accord with the concept that early hunters, not climate change, caused the extinctions of megafauna.

Recently Jackson and Weng (1999) challenged the view that all trees escaped extinction in near time. They reported an extinct species of spruce *(Picea critchfieldi)*, characterized by a seed cone of unusual size. Its late-Quaternary fossils are found in the southeastern states. Jackson and Weng have attributed its extinction to late-glacial climatic change, and those who believe that the same climatic impact accounts for megafaunal extinction (Grayson 2001) quickly picked up on this exception to the claim that plants suffered no extinction in the late Quaternary.

The simplest response to this argument is that the extinction of a single plant species presumably (but not clearly) coeval with large animal extinction is hardly enough to sweep away the overkill model. There is no reason to assume that climate changes generally tolerable to mammals would also have been tolerable to every species of plant in existence at the time. During the rapid climatic fluctuations of the late glacial and early Holocene, a temperate spruce species of limited range may have lagged behind its habitat, that is, dispersed more slowly than the climate changed.

On the other hand, might Critchfield's spruce have been eliminated by human activity? It should not surprise us if an occasional plant species were drawn into the overkill vortex. For example, when subjected to climatic stress, including drought, Critchfield's spruce would have been especially vulnerable to out-of-season anthropogenic ignitions, that is, the fires of spring. If so, one hypothesis would be that this temperate conifer fell victim to fire drives by the Clovis hunters—the same drives that may have favored fire-tolerant trees, such as ponderosa pine, as noted earlier. The conifer *Torreya*, a relict tree on the banks of the Apalachicola River in northern Florida and southern Georgia, may have had a history similar to that of Critchfield's spruce and barely escaped extinction (which may yet be its fate). When trees vulnerable to or favored by fire show striking changes in the fossil record around the time that people arrive in new lands, the possibility of an anthropogenic agency initiating, or at least furthering, the changes should be entertained. With the arrival of humans and the extinction of the megafauna, anthropogenic fire replaced herbivory by large mammals as the probable consumer of most above-ground savanna and grassland biomass. Fires can trigger the growth of human food plants in such environments and may have been set for this purpose.

As the 1960s drew to a close, we were equipped with several crucial new tools for exploring the late Quaternary and near time. Using fossil pollen analysis, packrat midden analysis, and radiocarbon dating, field teams from the Desert Laboratory proceeded to sample rock shelters and caves in the Grand Canyon and adjacent canyon lands. This provided rich opportunities to see what some of the large-animal extinctions looked like up close.

GIANT MEAT-EATING BATS?

I love tempting diversions. Not long after arriving in Tucson and setting up a pollen lab, I read news stories about a cable built across the Grand Canyon to a guano mine. On the Hualapai reservation at Quartermaster View, a few miles upstream from Rampart Cave, huge buckets suspended by the cable transported guano from Bat Cave across the canyon to fill trucks on the south side ("Treasure of Granite Gorge," *Time*, September 23, 1957). A *New York Times* headline on the story read, in part, "Big Vacuum Cleaner to Be Used to Mine Deposit Left by Giant, Meat-Eating Bats Millions of Years Ago" (March 27, 1957).

The headline is a hoot. I suppose that free-tailed bats *(Tadarida brasiliensis)* can be called meat eaters, since they eat insects. However, giants they are not, even among bats, nor was this vast guano deposit likely to be millions of years old. Nevertheless, the cave sounded interesting. The ground sloth skeleton exhibited at Yale's Peabody Museum, which was found at Aden Crater, New Mexico (together with long toenails, hair on a patch of skin over the rump, and dung balls) had been buried in bat guano (Lull 1930). Maybe Bat Cave sheltered its own ground sloth carcass. Furthermore, might bat guano not be as rich in fossil pollen as was ground sloth dung? Who knew what trophies might be found in desert caves, caves as dry as the tombs of the pharaohs? I simply had to see this operation for myself.

I admit that the Colorado River lacks pyramids, a sphinx, and a Valley of the Kings and Queens. But for all its archaeological treasures, even Egypt does not have perishable remains of any extinct animals as extraordinary and mysterious as the Shasta ground sloth. If not as hot and dry as Egypt, the lower Colorado River area is sufficiently arid to preserve a few remarkable mummies of its own. "Never mind the gold and buried treasure," I might have said, had I prayed to Ra, the Sun God, who has the head of a hawk and wears the solar disc as a crown. "I ask only for a previously unknown and undiscovered cave of layered sloth dung, with mummified ground sloth remains!"

In June 1958, geochronologist Bernie Arms and I followed a maze of dirt roads from Kingman, Arizona, to Quartermaster View, a splendid overlook into Grand Canyon. We parked near the tall cable tower and walked to the rim. Suddenly, there it all was, an amazing vista down to a tiny brown strip far below, with white streaks marking rapids in the muddy Colorado. The sagging cable faded from view. At 9,010 feet in length and 1.5 inches in diameter, this was reportedly the longest commercial cable in the world. U.S. Steel had reportedly spent $689,000 to manufacture it ("Treasure of Granite Gorge," *Time*, September 23, 1957). The expense seemed justified by the size of the deposit—an estimated 100,000 tons of guano (*New York Times*, March 27, 1957) expected to yield, according to the U.S. Guano Corporation's optimistic projection, a profit of $12.5 million.

The engineer in charge, Bill Freiday, arranged for us to cross the canyon the next day. When the time came, I assumed an indifference that I did not feel and followed Bernie into a head-height steel bucket, engineered to hold a load of

3,500 pounds. We swung out across the canyon, an adrenalin rush accompanying the giddy descent, and eventually landed on the north side of the Colorado, one mile away and 3,650 vertical feet below the rim, passing only one tower in between. Then there was a short cable car run, on what seemed like a sled, up to the huge mouth of the cave itself.

The miners seemed pleased to find someone, even greenhorn academics, interested in what they were doing. U.S. Guano Corporation chemist Varley Crompton had previously written me, "We occasionally run across bat 'grave yards' which are heaps of skulls and bones, sometimes up to four or five feet in diameter" (personal correspondence, March 17, 1958). The miners confirmed that there were plenty of bat bones and showed us mummified free-tailed bat carcasses. But they reported nothing large, nothing like a giant meat-eating bat, and certainly no big bones of ground sloths or other large animals.

We collected guano samples near the mouth of the cave and farther in, beyond an active bat colony, in a large, totally dark interior room filled with odorless dry fossil guano the texture of face powder. Our sample from a depth of just over 7 feet later yielded a date of 12,900 ± 1,500 radiocarbon years; the large error margin indicated that the sample contained too little organic carbon to yield a more precise measurement. Although we did not find any fossils of extinct animals, the deposits might well have been old enough to contain them; perhaps, under the unexcavated guano pile, they lie there still. The samples also contained no fossil pollen, though they did contain scales of small moths and fragments of beetle exoskeletons.

Exciting as the trip had been, Bat Cave was, for our purposes, a washout. Nevertheless, I was sure that once word of our interest in cave deposits began to spread, someone in the region would pass on news of previously unknown or unappreciated ground sloth caves like Rampart Cave. I had to wait twenty years, however, before Ra finally delivered, letting me join a team studying the dung and diet of the largest extinct mammalian megaherbivore known in Arizona.

Not long after our visit to Bat Cave, I learned that a military jet from a base in Nevada, apparently hot-dogging illegally within Grand Canyon air space, had nicked the cable. It was ruled unsafe. The U.S. Guano Corporation recovered its investment from the government and did not rebuild. I am told that the Hualapai tribe presently operates a casino at Quartermaster View, with clients ferried in from Las Vegas by helicopter.

FOUR

GROUND SLOTHS AT HOME

When a thing ceases to be a subject of controversy, it ceases
to be a subject of interest.

William Hazlitt, "The Spirit of Controversy"

Although I analyzed fossil pollen in sloth dung samples from Rampart
Cave in 1958, I did not actually visit the cave itself for another decade.
In the meantime I studied glacial-age vegetation change based on fossil
pollen records from Costa Rica, Arizona, California, New Mexico, and
Utah. I compared large-animal extinctions in East Africa with those in
Madagascar. Yet I never made the two-day trip from Tucson to Rampart
Cave. My walking was deteriorating and I elected to take longer trips
while I still could. By 1969 the time had finally come for a good look at
ground sloth dung *in situ*.

In January, with the help of National Park Service (NPS) rangers and
their river patrol boat, I found myself at the west end of the Grand
Canyon a few miles downstream from Bat Cave. Scrambling up to a ledge,
I caught up with my guides and let my eyes feast on the view. Any view
of the Grand Canyon, from the top down, the bottom up, or, like this,
from somewhere in between, is stunning. Five hundred feet below our
perch, the Colorado River flowed west toward a gap in the Grand Wash
Cliffs. The south-facing slopes and canyon walls were sun-baked and supported fewer woody plants than those facing north. On a steep slope to
the west of Rampart Cave, only lichens grew on an apron of rock rub-
ble detached from the cliff face above. The talus might once have provided retreats for yellow-bellied marmots, whose bones have been found
in Rampart and other caves in the canyon. In contrast, in winter the north-

facing slopes were relatively cool and moist. Near the cave we found a rich crop of shrubs and herbs. Occasionally plants found root space between or on top of great blocks of limestone, evidently once transported down the steep slope by mighty mudflows.

Inside the cave, my NPS guides unlocked a gate and returned downstream, reminding me to lock up when I left. Austin Long and the rest of our party, hiking in from Pierce Ferry, soon arrived. Slowly proceeding deeper into the cave, we fell silent as in a cathedral, thrilled to experience one of the most remarkable fossil deposits in the world. The back of the cave was faintly illuminated by light from the cave mouth. In single file we walked into a trench, through sloth dung. When we stopped we stood chest deep in layers of stratified sloth dung. There was no perceptible airflow, but the deposit had lost any trace of ammonia or other odors of decaying manure; the air smelled resinous, like incense. No one spoke a word. In the stillness I felt the hair rise on the back of my neck. One did not need to be a Sufi or a mystic to sense that this dimly lit, low-ceilinged chamber was a sacred sanctuary. More than a sepulcher for the dead, Rampart Cave venerated the extinct.

The trench walls were packed solid. When fresh, the 40-by-50-foot deposit must have been compressed by the repeated tromping of 500-pound ground sloths. Near the middle of the deposit a layer of plant debris brought in by packrats was interleaved with layers of the compacted dung. Around the edges where the deposit adjoined cave walls and in places on top of it, we found untrampled dung balls 4.5 to 6 inches in diameter, the size of softballs. These should have been the last to be deposited, and we collected some for radiocarbon dating. We sought the final record of a large mammal teetering on the threshold of extinction. We would discover similar opportunities in Stanton's Cave at the upper part of the Grand Canyon.

The dung was remarkably well preserved, not only at the upper levels but also down to at least 4.5 feet, where it dated at over 30,000 years old. One measure of the quality of preservation is the ratio of carbon to nitrogen, which changes rapidly with exposure to the elements. The carbon-to-nitrogen ratio of the ground sloth dung was not appreciably different from that of fresh cow manure (Clark, O'Deen, and Belau 1974). To be certain that the dung was much more than a few hundred years old, we obtained radiocarbon dates on sloth dung from the top to the bottom of the deposit.

From earlier reports we had every reason to expect that Rampart Cave would yield riches, for it had been explored for fossil bones long before

Plate 2. View upstream from mouth of Rampart Cave, April 1971. Photo by Jim King.

Dick Shutler took his samples in the 1950s. I believe that Willis Evans was the first to recognize its paleontological significance. In the mid-1930s Evans, a Pit River Indian from Northern California (Harrington 1933), led a site survey crew that mapped and excavated small shallow caves in the western end of the Grand Canyon. A few years earlier, Evans had helped anthropologist Mark Harrington excavate Gypsum Cave, Nevada, which harbored an extraordinary deposit of ground sloth dung along with bones of various extinct animals and prehistoric artifacts. As a result of his work in Gypsum Cave, Willis Evans knew what he had found when he entered Rampart Cave. It is fair to say that he knew his shit.

In two test pits in a bone-rich deposit in the back of Rampart Cave, Evans excavated through packrat middens and sloth dung to bedrock. The pits yielded fossil bones of Shasta ground sloth, yellow-bellied marmot, jackrabbit, ring-tailed cat, bobcat, an extinct mountain goat *(Oreamnos harringtoni)* first collected by Harrington in Nevada, an extinct burro-sized horse, and possibly mountain lion. In addition there were bones of gopher tortoise, chuckwalla, and unidentified birds, later recognized to include California Condors. Some of the fossil bones had tissue attached. There was also hair, apparently that of the ground sloth (Harington 1972; Harrington 1936; R. Wilson 1942). (Dick Harington,

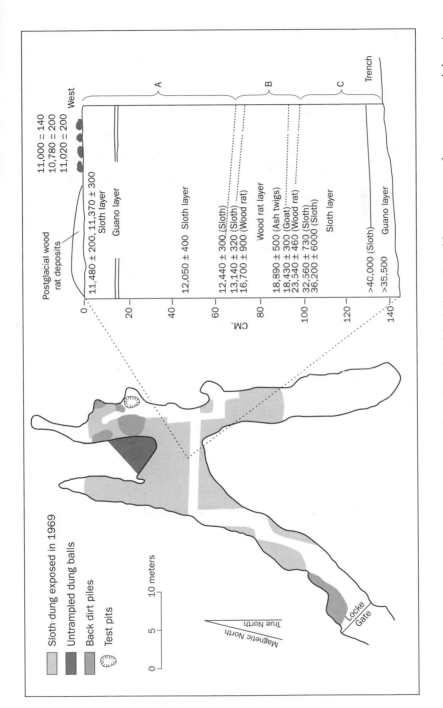

Figure 11. Floor plan of Rampart Cave, Arizona, and stratigraphy of sloth dung. Roughly 95 percent of an unexcavated deposit was destroyed by fire in 1976. Adapted from Long and Martin 1974. Used with permission from *Science*, © 1974 AAAS.

Plate 3. "Dance floor" of sloth dung in Rampart Cave, April 1971. In 1976, the deposit was destroyed by fire. Photo by Eugene Griffen.

not to be confused with Mark Harrington, is a Canadian paleontologist and mountain goat expert who joined one of our Rampart Cave trips.)

In 1942 Remington Kellogg and his crew screened a large volume of sloth dung and packrat middens at Rampart Cave. They found only a few more species to add to Evans's collection of fossil vertebrates. In addition, archaeologist Gordon C. Baldwin, detailed to look for any indications of early humans, wrote, "There was not a single fragment of evidence to indicate that man had ever occupied the cave, either contemporaneously with the ground sloth or later" (Baldwin 1946). The oldest artifact that we found on our trip in 1969 was a bit of newspaper from the 1930s. A small headline read, "Fascists Bomb Mallorca."

Our goal in 1969 was to clarify an ambiguity resulting from the pilot study that I had published with Dick Shutler and Bruno Sabels a decade earlier. In assessing the overkill theory (particularly any version of it that involved rapid extinctions), it was vital to ascertain as precisely as possible both when humans arrived on the continent and how long thereafter the large mammals survived. The available radiocarbon dates indicated that the Clovis people, the First Americans, had reached this part of the continent approximately 11,000 radiocarbon or 13,000 calendar years ago. The youngest of Shutler's dates, based on both plant residues

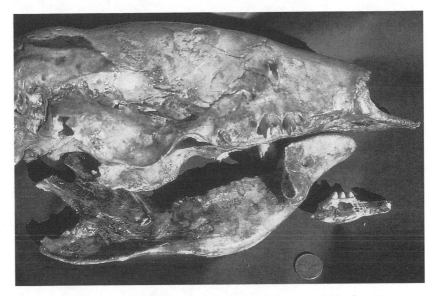

Plate 4. Shasta ground sloth adult cranium and infant mandible, from Rampart Cave, with penny for scale. Photo by author of specimens in the National Park Service Collection, Grand Canyon National Park.

and humic acids in the dung, indicated that sloths had lived at Rampart Cave until 10,000 radiocarbon years ago. Recent findings, however, had begun to cast doubt on this and other late dates, as well as on much younger dates on various extinct large animals generated in the first erratic years of radiocarbon dating. Libby, for example, had obtained an 8,500-year date (lab catalogue number C-222) on sloth dung from Gypsum Cave. None of our ground sloth samples from caves in Arizona, Nevada, New Mexico, and Texas were as young (see table 5).

Now armed with improved methods, investigators began to challenge all Holocene dates on extinct animals. In southern Arizona, Vance Haynes found that mammoths associated with human artifacts in a kill or processing site were all of Clovis age, about 13,000 calendar years old, and that mammoth extinction had occurred closer to 11,000 radiocarbon years ago than the 8,000 years or less that the first Arizona radiocarbon dates suggested. Therefore, if Shutler's 10,000-year date, L-473A, at Rampart Cave was valid, the Shasta ground sloths had disappeared a thousand years after the mammoths. This was a puzzling discrepancy, as the ground sloths should have been, if anything, more vulnerable to hunters. They were ambulatory pin cushions, helpless beasts unless a predator came close enough to be ripped with those long claws at the ends

Plate 5. Stratified sloth dung and packrat middens, Rampart Cave, ca. 1969 or 1970. Photo by the author.

of its long arms. If people and ground sloths had coexisted for a thousand years, my version of the overkill theory was in trouble.

We obtained radiocarbon dates on 14 samples, collected mainly on the surface of the dung deposit. As shown in table 5, all but one of our dates were hundreds of years older than Shutler's. Our youngest date, A-1067, was 10,780 ± 200 radiocarbon years. The discrepancy could have meant that the original measurement was in error, either because dating technology had not been sufficiently refined or because the sample was contaminated. It could also have meant that we simply had not dated enough young dung balls the second time around to reflect the true sample range.

Much depended on an accurate determination of when the last dung was deposited, so we pursued all of these possibilities. We hunted down other deposits of ground sloth dung for further tests. After three decades we made another test. Pete Van de Water located and cleaned a fragment

TABLE 5 *Radiocarbon Samples from Ground Sloth Caves*

(solid carbon dates excluded)

Site and sample no.	Material	Location in cave	Depth (cm)[a]	Laboratory no.	^{14}C date (years ago)
Rampart, AZ					
1	Sloth dung ball	Surface		A-1066	11,000 ± 140
2		Surface		A-1067	10,780 ± 200
3		Surface		A-1068	11,020 ± 200
4	Ephedra twig	Surface	0–5	CAMS-19997	10,940 ± 60 (from L-473A)
5	Trampled sloth dung	Surface	0–5	A-1392	11,370 ± 300
6		Surface	0–5	A-1041	11,480 ± 200
7		Surface	0–5	L-473A	10,035 ± 250
8		A	46	L-473C	12,050 ± 400
9		Base of A	61	A-1070	12,440 ± 300
10	Sloth dung ball	Unknown		A-1318	12,470 ± 170
11	Sloth dung	Top of B	67	A-1207	13,140 ± 320
12	Packrat pellets	B	71	A-1208	16,700 ± 900
13	Twigs of ash (*Fraxinus*)	B	90	A-1356	18,890 ± 500
14	Goat dung	B	91	A-1278	18,430 ± 300
15	Packrat pellets	Base of B	96	A-1209	23,540 ± 460
16	Sloth dung	Top of C	99	A-1210	32,560 ± 730
17		Top of C	99	A-1043	36,200 ± 6,000
19		Base of C	132	A-1042	>40,000
19	Bat guano	Base of C	137	L-473D	>35,500
Muav, AZ					
20	Sloth dung	Surface		A-1212	11,140 ± 160
21		Surface		A-1213	11,290 ± 170
Gypsum, NV					
22	Sloth dung	Room 3		LJ-452	11,690 ± 250
23		Unknown		A-1202	11,360 ± 260
24		Unknown		L/a-11835	19,875 ± 215

TABLE 5 *continued*

Site and sample no.	Material	Location in cave	Depth (cm)[a]	Laboratory no.	[14]C date (years ago)
Aden Crater, NM					
25	Sloth dung			Y-1163B	11,080 ± 200
26	Body tissue[b]			Y-1163A	9,840 ± 160
La Gruta del Indio, Argentina					
27	Sloth dung	70–80 cm,		A-1351	10,740 ± 150
28		80–90 cm,		A-1371	11,350 ± 180
29		1.10 cm,		GRW-5558	10,950 ± 60
30		70 cm, R8		A-1370	24,730 ± 860
Cueva del Milodón, Chile					
31	Sloth dung	Unstratified		A-1390	13,560 ± 190
32	Hair and skin	Unstratified		R-4299	13,500 ± 410
33	Hide	Unstratified		A-1391	10,400 ± 330
34	Dung	Unstratified		SA-49	10,200 ± 400

[a]All depths listed are from same vertical profile. Values are midpoints of 3-centimeter depth ranges.
[b]Discrepancy with Y-1163B perhaps due to organic preservative. For more Cueva del Milodón dates, see Markgraf 1985.

of Mormon tea *(Ephedra)* twig in one of Shutler's surface samples. The specimen was submitted to Thomas W. Stafford's Laboratory for Accelerator Radiocarbon Research, then in the Institute of Arctic and Alpine Research at the University of Colorado at Boulder. In a letter dated May 15, 1995, we received the following result from Stafford: "14C AGE— 10,940 ± 60 (CAMS-19997)." This meant there was a 95 percent chance that the age of the specimen fell between 10,820 and 11,060 radiocarbon years before the present, within the range of surface dates obtained in our second round (Long and Martin 1974). Possibly the original date (L-473A; see table 5) had incorporated a mixture of materials, with a small amount of postextinction packrat debris contaminating the youngest sloth dung.

The oldest of our 1969 results also suggested that the sloths first entered Rampart Cave more than 40,000 years ago. For some unknown reason, possibly high water in the Colorado River, they left around 32,000 years ago, abandoning the cave to packrats, marmots, mountain goats, and ring-tailed cats. Twenty thousand years later, the sloths returned. Then, judging by the rate of dung deposition between 13,000 and 11,000 radiocarbon years ago, they flourished right up to the time when deposition ceased. About 10,000 to 11,000 radiocarbon years ago, the Shasta ground sloth seems to have died out, not only at Rampart Cave and adjacent Muav Cave, but throughout its range, which included populations separated both geographically (Nevada to West Texas) and ecologically (the Mojave Desert to spruce woodland) (Martin, Thompson, and Long 1985).

One evening I perched at the mouth of Rampart Cave in a meditative mood. Over 12,000 years ago a Shasta ground sloth might have done the same. The sun sank behind canyon walls, enhancing a sense of sanctity. Shadows lengthened. Breezes of the day died down. In the stillness, the cool air of an early winter evening settled into hollows. I imagine myself as a ground sloth at the moment of arrival of the first people. I am lounging at the mouth of my cave. I have never seen these strange, two-legged creatures before and I have no knowledge of the danger they pose. To avoid cactus they lift their feet high as they stride along, like great blue herons stalking frogs. Now they see me, having discovered my spoor at the bottom of the slope. They raise their arms; every hand holds a rock or a club. I am defenseless against them, and this is the last thing I see.

We also used the Rampart Cave dung samples to further elucidate the ground sloth's diet, which soon became the best-known aspect of its ecology. (Extinct parasites were also found in the dung; Schmidt, Duszynski, and Martin 1992.) Shutler's samples had contained not only globe mallow and juniper pollen (see chapter 3), but also large amounts of pollen from Asteraceae (plants in the sunflower family). Eames (1930) had reported Asteraceae fragments in ground sloth dung at Aden Crater, New Mexico (see Aden Crater dates in table 5). But we wanted more detail. It was Norrie (Eleanora) Robbins, a Desert Lab student working on the diet of the extinct goat, who told me about Dick (Richard M.) Hansen and his composition analysis lab at Colorado State University. Dick investigated the composition of rumen or dung samples from cattle, bison, and other large herbivores, an excellent index to their diet. His method of particle analysis was an elegant technique for determining the forage consumption of living large herbivores. His laboratory maintained reference collections of native forage plants of the western region. He had trained staff who could identify fragments from the digestive tract or excreta of wild or domestic herbivores. Dick agreed to try his method on a sample of ground sloth dung. To our delight, we learned that the dung samples contained well-preserved and identifiable plant parts. We joined forces.

At Rampart Cave Dick collected 514 dung samples from appropriate units in the walls of the trench dug by Kellogg. Back in his lab he supervised microscopic analysis of plant parts recovered from samples of various ages. The lab confirmed that although the ground sloths might have ingested pollen of wind-pollinated species such as juniper, flower heads of creosote bush, and flowers or pollen of Asteraceae, they did not ingest much foliage of juniper or other oily plants (Hansen 1978). However, ground sloth dung collected at over 6,000 feet in the Guadalupe Mountains of West Texas (Van Devender and others 1977; Spaulding and Martin 1979) revealed, somewhat to my surprise, a diet of conifer needles (Douglas fir, genus *Pseudotsuga*). So did samples, potentially of ground sloth dung, from Cowboy Cave in Utah. These records suggest a wider range of foraging than those at Rampart Cave. They also demonstrate the glacial-age elevational descent of Douglas fir, which at present does not grow in the immediate vicinity of either Cowboy Cave or the sloth caves in the Guadalupe Mountains.

Further confirming our earlier fossil pollen analysis (Martin, Sabels, and Shutler 1961), Dick found that desert globe mallow *(Sphaeralcea laxa)* leaves or stem fragments made up on average half of the plant parts found in ground sloth dung at Rampart Cave (Hansen 1978). In addition, composition analysis indicated an average diet of 18 percent Mormon tea *(Ephedra)*, with its distinctive, many-furrowed pollen grains; 7 percent saltbush *(Atriplex)*; 6 percent catclaw acacia *(Acacia greggii)*; 5 percent common reed (a tall aquatic grass, *Phragmites*); 2 percent yucca *(Yucca* sp.); and lesser amounts of other succulents, including cacti. Dick's laboratory also identified 65 other plant genera, presumably minor dietary items from "accidental bites." Except for *Phragmites*, Dick found no more than traces of grasses in the dung; these could have been "accidentals" (Hansen 1978).

Data from Rampart Cave and its environs subsequently generated more information on the ground sloth's diet and possible behavior. More generally, it expanded our understanding of the ecology of the region in glacial times. For example, from fossil midden contents, Art Phillips (1984) confirmed that the ground sloths relied heavily on globe mallow.

From contemporary packrat middens he also discovered that when ground sloths roamed the western end of the canyon, cool-adapted plant species were more numerous than at present. For example, juniper and single-leaf ash *(Fraxinus anomala)* are common plant fossils in packrat middens, indicating that junipers and ash trees or shrubs once grew near the cave mouth. Now juniper and ash grow at higher elevations or in wetter sites, and near the mouth of the cave one finds only the shrubs and cacti of Mojave desertscrub.

Recent work on a ground sloth dung ball from Gypsum Cave by Hendrik Poinar and other members of Svante Pääbo's team in Leipzig, Germany, yielded not only Shasta ground sloth DNA but also DNA from plant remains incorporated in its diet (Poinar and others 1998). Their findings indicated that the ground sloths foraged not only in drier sites but also in riparian communities supporting wild grape *(Vitis)* and very likely the large mustard *Stanleya*.

These results raise interesting questions as to the time of year the ground sloths spent in the cave. From eastern California to West Texas, ground sloth bones, unlike those of mammoths, camels, and extinct bison, are found more often in caves than in midvalley floodplain alluvium or lake deposits. Despite their rich fossil record in caves, however, I doubt that the sloths spent much time in Rampart, or they soon would have filled it with dung. We estimated that the average annual rate of deposition in the upper part of the deposit was slightly more than a cubic foot a year, an amount that probably represented less than a week's elimination from one healthy adult ground sloth. The presence of embryonic sloth bones suggests that females may have used Rampart Cave as a nursery. Bones of a baby ground sloth the size of a cat found in Gypsum Cave (Harrington 1933, 78) point to a similar conclusion there. This makes particularly good sense if the infant sloths, like many mammals, were born in the spring. I believe late winter or spring following a wet winter would have been the best time for the ground sloths to find forage at low elevations in a xeric habitat such as that surrounding Rampart Cave.

As spring ended and globe mallows and other forage plants dried and shriveled, I expect the ground sloths left the inner gorge for greener pastures at higher and cooler elevations. The nearest place to summer above 5,000 feet in a tolerable climate would be in the Grapevine Mountains and the Garnet Mountains of the Grand Wash Cliffs, within about 40 miles of Rampart Cave. Projecting from late-glacial plant displacements recorded elsewhere (see chapter 3), I would expect that above the juniper–single-leaf ash woodland at Rampart Cave, the sloths would have penetrated a pinyon-pine woodland and perhaps even reached a few limber pine or Douglas fir trees, with sagebrush *(Artemisia tridentata)* on the flats. Because of the glacial age climate, the vegetation gradient not only from the Grand Wash Cliffs to Rampart Cave but throughout the West would have been less arid and more productive in plant dry matter from forage plants than it is now.

At a very leisurely sloth travel rate averaging one to two miles per day, the vertical migration I propose from canyon bottom past Mead View and up into the Grand Wash Cliffs need have taken no more than a month or two. In the fall, as the weather turned chilly, the ground sloths would slowly have found their way back down again to lower and warmer elevations, perhaps drawn by plants growing along the Colorado during the winter season of low water. In short, I am betting that the ground sloths of the Quaternary, like modern-day elk and mule deer, took advantage of elevational gradients to benefit from seasonal changes in availability of forage.

Finally, the dietary data from ground sloth dung help refute the argument that its extinction around Clovis time resulted from climate change. The ground sloth's favorite food plants—in the Rampart and Muav cave areas, Nevada's Gypsum Cave, the sloth caves of the Guadalupe Mountains of West Texas, and New Mexico's Aden Crater—remain important components in the vegetation of arid regions in North America.

Thanks especially to the remarkable stratified record from Rampart Cave, we know that for thousands of years the Shasta ground sloth browsed on a variety of desertscrub and woodland shrubs and forbs, including species presently favored by wild desert bighorn sheep and feral burros. Even the fossils indicating that the ground sloth once lived in what is now Sonoran Desert—archaeologists have excavated its bones from Ventana Cave on the O'odham reservation west of Tucson—appear to coincide in time with the invasion of many Mojave desertscrub species preserved in the fossil packrat middens at Organ Pipe Cactus National Monument. Most of the plants identified in Rampart Cave sloth dung (Hansen 1978) remain important in the natural vegetation within the seasonal range of the Grand Canyon ground sloths. It is therefore not obvious how one might account for the Shasta ground sloth's extinction by invoking a loss of food supply resulting from climate change. Single-leaf ash and juniper continued to grow near Rampart Cave for at least 2,000 years after the ground sloths disappeared (Phillips 1984). Though these species were not food sources for the sloths, their persistence confirms that no major climate change is likely to have disrupted the animals' food supply (Martin 1986, 122).

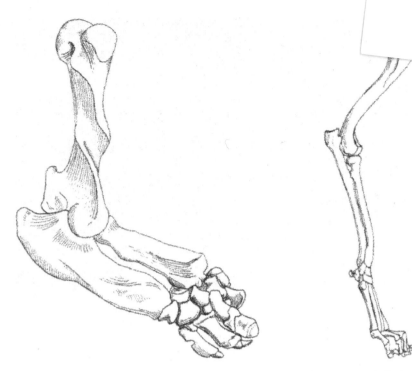

Figure 12. Comparison of hind legs of the slow-moving Shasta ground sloth and the fleet coyote. Reprinted from Kurtén 1988, © Columbia University Press.

Not dreaming what the future would bring, in 1973 Austin Long and I guided author James Michener through the cave, and he wrote about it in *Reader's Digest* (see plate 6). Michener understood that this was one of only a dozen caves known to contain fossil sloth dung, and that it held the best-preserved and deepest deposit of stratified sloth dung that has been found in North America. It deserved nomination as one of the wonders of the paleontological world, along with the frozen mammoths of Siberia, the thousands of tar-impregnated saber-toothed cats and dire wolf bones of Rancho La Brea, and the fifty young male mammoths and the giant short-faced bear in a sinkhole at Hot Springs, South Dakota.

In July 1976, Austin and I got a call from Roy Johnson at Grand Canyon National Park. The news was bad. He reported that smoke was coming from the mouth of Rampart Cave. We chartered a plane and flew to the canyon, where we joined a Park Service fire crew and rode by heli-

Plate 6. Author James Michener (white shirt) and National Park Service guides at the mouth of Rampart Cave, December 1973. Photo by the author.

copter to the cave. Heavy smoke creeping along at ground level poured slowly out of its mouth. The crew began to douse the fire. Then a large rock dropped from the low ceiling, perhaps loosened by the heat. More cautious measures were adopted. Eventually the cave ceiling was shored up with massive posts in an effort to make it safe for firefighters to enter. (Perhaps inevitably, a columnist in the *Chicago Sun-Times* ridiculed the National Park Service for spending tens of thousands of dollars in an unsuccessful effort at saving ancient dung. Even renowned TV anchorman Walter Cronkite could not resist closing an interview with me on this paleontological disaster with a quip about "endangered feces.")

The dung smoldered for months, defeating all efforts at extinguishment. Although there was no real blaze—no visible flames—the insidious combustion slowly and inexorably reduced to ash the magnificent 5-foot-thick blanket of dung east of Kellogg's trench, the very area we had found so valuable in our research. While part of the deposit remains, the main portion, shown in figure 11 and plate 3, is gone. Too late, I regretted that our sampling had not been more intensive. Before more than a few intriguing studies had been completed, the fire all but closed a precious window into the late-Quaternary ecology of the Grand Canyon ground sloths. In its own way, the Rampart Cave fire was as destructive

Plate 7. Smoke from fire, mouth of Rampart Cave, July 1976. Photo by the author.

of information as the long-lamented conflagration of the ancient library in Alexandria, Egypt. Had we given the cave too much publicity? Whatever the explanation, I continue to mourn the loss.

For more information on the extinction of ground sloths throughout their hemispheric range, we needed to look beyond the Southwest, and indeed beyond North America. I had a chance to do just that in 1972, thanks to a sabbatical from the University of Arizona, an NSF grant to cover research costs, and the interest of Ike and Jean Russell, adventurous friends in Tucson who let me fly with them in their Cessna through Central and South America. We headed for southern Chile and the famous Cueva del Milodón (Mylodon Cave, also known as Eberhardt Cave).

At Seno Ultima Esperanza (Last Hope Sound), an arm of the sea behind Andean glaciers near Puerto Natales, settlers over a century ago had learned of this grotto, 400 feet wide, 100 feet high, and 660 feet from front to back. It was the first cave found to contain ground sloth dung— along with hair and, most remarkably, a piece of hide 3 feet in diameter, patched with hair and embedded with dermal bones. Despite its youthful appearance, radiocarbon dates obtained from this prize specimen are approximately 13,000 years old (A-1390, R-4299).

Based in part on dung from this cave, the famous Argentinean pale-ontologist Florentino Ameghino described what he imagined was a new genus and species of living ground sloth, *Neomylodon listai*. Other authorities called it *Grypotherium domesticum* (Hauthal, Roth, and Lehmann-Nitsche 1899). Their work was consistent with contempora-neous travelers' belief that ground sloths still lived in southern Chile. There were legends of a strange animal known not only to the Indians but also to European explorers. One hundred years ago a London news-paper financed an expedition to search Patagonia for living ground sloths. The scientific world was ablaze with curiosity. Alas, the search failed.

Although disturbed by various excavations, Cueva del Milodón still contains an unrivaled deposit of sloth dung along with fossil remains of extinct horses (*Hippidium, Onohippidium*) (Latorre 1998; see Sutcliffe 1985 for a splendid illustration of the cave). On our 1972 trip we col-lected surface or shallowly buried, and therefore potentially the youngest, samples of what we took to be ground sloth dung for radiocarbon dat-ing. Possibly some of the samples came from extinct horses, not from mylodons. Judging from the content of their enormous dung balls, in some cases larger than a circus elephant's, ground sloths from Cueva del Milodón ate mainly grasses.

To top off the trip, we visited another cave yielding not only sloth dung but also archaeological remains. Called La Gruta del Indio (Indian Cave) and located on the Río Atuel, outside the small city of San Rafael at the foot of the Andes in Argentina, it had recently been reported by Tito (Hum-berto) Lagiglia, director of San Rafael's Museum of Natural History. Tito and his family made us most welcome in San Rafael, and he escorted us to the cave (as well as accompanying us to Cueva del Milodón).

La Gruta del Indio is at a latitude equivalent to that of Las Vegas and is surrounded by similar habitat. Although lacking Joshua trees, the veg-etation resembles in structure, height, density, and spacing of dry land shrubs the desertscrub near Rampart Cave and in other parts of the Mo-jave Desert. Lush mountain grasslands occupy higher elevations, with alpine plants in the Andes. The cave itself, actually a rock shelter beneath a basalt ledge, is smaller and much more exposed than Rampart Cave. Like Rampart, it harbors middens of some plant-gathering, packrat-like mammal. Unlike Rampart, La Gruta del Indio is an archaeological site, with charcoal deposited immediately above the layers harboring the youngest dung balls (Long, Martin, and Lagiglia 1998) and with late-prehistoric artifacts.

La Gruta del Indio offered very few bones to help with species identification. The discovery of one megatheriid tooth and of dermal ossicles embedded in patches of hide (the mark of a mylodon) suggested that both mylodons and megatheriids had been present. The ground sloth dung balls at La Gruta del Indio (see plate 8) were two to four times smaller than those found in Rampart. (I was intrigued to learn from the study of Hector d'Antoni [1983] that the Argentinean ground sloths ate mesquite pods, which we had not found in the dung of the Shasta ground sloth.)

A core question for me throughout this trip was whether the youngest ground sloth dung deposits in South America were the same age as those in North America. An answer could help us ascertain whether the two groups of megafauna had gone extinct at the same time. One previously published radiocarbon date on hide from the Groningen Laboratory, 9,560 ± 60 (GrN-5772), was decidedly younger than any dates that Austin Long and I had obtained on the Shasta ground sloth. Could ground sloths in southern South America have lasted 1,000 to 2,000 years longer than those in North America?

Almost all the ground sloth boñegas (dung balls) from La Gruta del Indio (Long, Martin, and Lagiglia 1998) and Cueva del Milodón (Markgraf 1985) (see table 5) are over 10,000 radiocarbon years old. Those from Cueva del Milodón are no older than 14,000 radiocarbon years. Alejandro Garcia and Tito Lagiglia (1999) recovered much older dates at La Gruta del Indio. With the exception of an early University of Chicago date on mylodon dung that I believe is in error and should be discarded, the youngest dates on the samples from La Gruta del Indio and those Lagiglia and I collected from Cueva del Milodón are quite similar. More than that, they are similar to the surface dates on the Shasta ground sloth dung from Arizona. Based on these dates, if the First Americans caused the extinctions of ground sloths, they spread very rapidly from Arizona to southern South America, so rapidly that some archaeologists question whether it was possible.

A more recent series of six radiocarbon dates from La Gruta del Indio, however (Garcia and Lagiglia 1999; Garcia 2003), includes one that matches the earlier Groningen date. Both are younger than any reported in Long, Martin, and Lagiglia 1998, and in fact any radiocarbon dates on ground sloth dung and other samples of high quality reported in recent years. Thus the question of whether ground sloth extinction in southern South America significantly postdated extinctions in North America is still open.

Though younger age estimates appear from time to time (see Sutcliffe 1985 on work by Saxon in Cueva del Milodón), no ground sloth remains

Plate 8. Mylodon dung from La Gruta del Indio, Argentina, collected by Humberto Lagiglia. Museo de Historio Natural, San Rafael. Photo by the author.

in North America have been discovered in reliably dated geologic deposits of the last 10,000 years. With the crucial exception of the West Indies, ground sloth remains are absent from numerous Holocene fossil deposits, as are the rest of the near-time extinct megafauna of North and South America. Continental deposits give no hint of a later survival. So matters stood until dating results came in from Cuba and Haiti. There dwarf ground sloths persisted until about 5,000 radiocarbon years ago, approximately the time of settlement (Steadman n.d.).

The South American ground sloth investigations continue. Recently Michael Hofreiter and others (2003) reported a sloth dung cave at the foot of the Andes at 38.5 degrees south. A particle accelerator yielded an age of 14,665 ± 150. The investigators had the benefit of mitochondrial DNA analysis. This technique for analyzing cave earth, cave coprolite, and subarctic frozen ground is a recent advance in detecting the presence, the identity, and even the diet of extinct animals (Poinar and others 1998, 2003). On the basis of mitochondrial DNA, Hofreiter and others conclude that an undescribed small species of ground sloth lived during the late glacial in the lower parts of the eastern Andes. It may be the same species that left the dung deposit at La Gruta del Indio.

One more cave dry enough to preserve fossil ground sloth dung deserves mention here. I would not have imagined such a find in the Brazilian tropics. Nevertheless, from Gruta de Brejões in Bahia, one of the drier parts of Brazil, Nick Czaplewski and Castor Cartelle (1998) obtained a date of 12,200 ± 120 (Rafter Radiocarbon Laboratory, NZA-6984) on dung associated with a skeleton of the small ground sloth *Nothrotherium maquinense*. In addition, Czaplewski (letter of December 19, 2000) called my attention to a study of plant remains in coprolite of an extinct Brazilian llama, *Paleolama*. These finds give us reason to hope that the caves of Bahia will also yield ample bone collagen suitable for reliable radiocarbon dates and DNA determination. This would constitute a major advance in refining the near-time extinction chronology of the New World tropics, rarely thought of as a suitable environment for preservation of perishable material.

Ground sloths also inhabited the West Indies, but here their story was very different. Zoogeographers dispute whether they got to the islands by swimming (living tree sloths are good swimmers) or by walking over dry-land connections with the mainland sometime during the Tertiary (Iturralde-Vinent and MacPhee 1999). In any event, in the Quaternary the West Indies harbored nine poorly known endemic genera: seven in Cuba, two in Hispaniola (one shared with Cuba), one in Puerto Rico (shared with Cuba), and one in Curaçao. All were in the family Megalonychidae. As one would expect of an ancient oceanic island fauna, many were severely dwarfed in comparison with their continental ancestors: with one exception all weighed less than 45 kilograms (100 pounds). Until their extinction they appear to have been the predominant mammalian herbivores, the largest plant eaters in the Greater Antilles, despite their dwarf size among the ground sloths.

Most interestingly from my perspective, robust fossil evidence suggests that the ground sloths survived into the Holocene in Cuba and Hispaniola. From a cave in Cuba, Ross MacPhee and others report a date of 6,250 ± 50 on the largest of the extinct dwarf West Indian ground sloths, *Megalocnus rodens* (MacPhee, Flemming, and Lunde 1999). A second date, presumably from the same locality, is 6,330 ± 50 (Beta-115697). A third, recently received from the accelerator lab of the University of Arizona, is even younger, 4,486 ± 39 (AA-58430). MacPhee's youngest date on another dwarf ground sloth species, *Parocnus brownii*, is 4,960 ± 280 (AA-35290) . Although we lack direct radiocarbon dates on all 16 genera of extinct ground sloths from North and South America, or on all eight under 45 kilograms from the West Indies, the early returns indicate that West Indian ground sloths lasted at least 3,000 years longer than the continental ground sloths. The results support an anthropogenic extinction model.

CRYPTOZOOLOGY, GROUND SLOTHS, AND MAPINGUARI NATIONAL PARK

Most zoologists suppress any dreams they may have of the survival of late-Quaternary extinct beasts, such as ground sloths living in some remote corner of Brazil. Such healthy skepticism has not deterred cryptozoologists ("cryptos") from organizing and funding searches, some well supplied with advanced technological equipment, for animals that paleontologists say are extinct or (in the case of Sasquatch and the Loch Ness monster) never existed. Press releases from these expeditions make great copy, and their,leaders are likely to appear on television. Like the public at large, cryptos are drawn to charismatic megafauna much more than to small creatures such as arthropods. At least I have yet to hear of enthusiastic searches for bizarre and entirely imaginary invertebrates like flying millipeds or singing earthworms.

David Oren, a Harvard-trained ornithologist now with the Emilio Goeldi Museum in Belém, Brazil, has claimed to have evidence of living ground sloths (Oren 1993). The original reports of these creatures, known to Indians and local hunters as *mapinguari,* came from the Amazon Basin on the Tapajos River. Whether or not Oren knew about the search for living ground sloths a century earlier in Tierra del Fuego, history was repeating itself. Along with paleontologist and ground sloth expert Greg McDonald of the U.S. National Park Service, Oren appeared on the Discovery Channel on the proper tributary of the upper Amazon, the alleged habitat of the *mapinguari.* What would they find?

Oren did his best to lure a *mapinguari* by imitating what its call was supposed to sound like. There was no response. An Amazonian Indian guide found a dung sample suspected to be from the *mapinguari.* According to McDonald, its DNA, extracted in Svante Pääbo's lab in Germany, matched that of *Tamandua,* the living arboreal anteater of tropical America.

My astute cryptozoological friend Richard Greenwell, who is properly concerned with scientific methodology, says that a scientist cannot exclude chance, whether one in ten or one in a million or one in a number approaching infinity. If the only evidence is negative, there has to be a possibility, however small, that *mapinguari* are still living. Richard likes to remind me that the coelacanth *Latimeria,* unknown in the fossil record since the Cretaceous, turned up alive in the haul of fishermen in the Indian Ocean near Madagascar in 1938. And in the 1990s, Laotian and Vietnamese zoologists (not cryptos) described two genera of previously unknown and unexpected large animals from west-central Vietnam and Laos: *Pseudoryx,* a bovid, and *Megamuntiacus,* a cervid, each weighing about 100 kilograms (220 pounds). Nevertheless, while the world would be thrilled at the fabulous discovery of living ground sloths, I do not give their seekers a chance.

Would I love to be proved wrong? Yes, indeed! One side of me is wholeheartedly rooting for David Oren. What a thrill it would be for me to see a living, breathing ground sloth in the flesh. More than that, their discovery in the Amazon would fuel the initiative needed to set aside one or more large reserves in the world's largest rain forest (Holloway 1999). The ongoing global destruction of rain forests is one of the unspeakable tragedies of our time.

As wonderful as it would be to find ground sloths alive, however, the sad fact is that these amazing animals have been absent from the fossil record for thousands of years. The mantra of paleontologists, "The absence of evidence is not evidence of absence," is popular among cryptozoologists. Nevertheless, only in the West Indies have ground sloths been found in the fossil record of the Holocene, and they surely are extinct. Thousands of years of absence from the fossil record of North and South America of an animal as obvious as a ground sloth, not a crossopterygean fish lurking in the depths of the sea, is a big step in the direction of incredibility. So many late-Holocene fossil deposits— thousands, if not tens of thousands—have been sampled or excavated in continental North and South America without yielding ground sloth remains, or those of any other large extinct Quaternary mammal, that it seems impossible that West Indies ground sloths are still alive. Alas, David Oren and others hunting for the *mapinguari* started their quest thousands of years too late. That should not detract from the goal of establishing a vast *mapinguari* tropical forest reserve to represent the likely habitat of extinct ground sloths, gomphotheres, and glyptodonts.

FIVE

GRAND CANYON SUITE

Mountain Goats, Condors, Equids, and Mammoths

We must never underestimate the patience of extinction.

Michael Rosenzweig, Win-Win Ecology:
How the Earth's Species Can Survive in the Midst
of Human Enterprise

"Dammit, Bob! What are you dumping?" I was pissed. It was early in the morning, not my best time of day—and in June on Arizona standard time, mornings come very early. I was talking to Robert C. Euler, professor of anthropology at Prescott College, then Arizona's only private four-year nondenominational college. Wayne Learn, Bob's helicopter pilot, had just unloaded me and my gear on the floor of Marble Canyon, next to the surging Colorado River, 100 feet below the mouth of Stanton's Cave and 2,000 feet below the canyon rim. The view of the layered cliffs is spectacular, top down or bottom up, and I would see it both ways.

But on this June 1969 morning Bob and I were climbing up over talus littered with fresh "back dirt," the screenings discarded in the process of separating what would be bagged and saved by the field team. I eyed the back dirt with dismay. I had heard that archaeologists excavating desert caves for artifacts left the dung of extinct animals behind. Now I had caught one red-handed. The back dirt included rocks and pieces of driftwood, but it was mainly composed of ancient fecal pellets. Many were relatively large; at over half a gram in weight, they would prove to be more than twice the dry weight of those of mountain sheep or deer, but smaller than those of elk.

Appalled and outraged, I informed Bob in no uncertain terms that the pellets might well be those of Harrington's extinct mountain goat, which we (and Remington Kellogg before us) had also found in Rampart Cave. He was throwing away invaluable information. The samples would be ideal for radiocarbon dating and dietary analysis. "This is good shit!" I moaned. "You can't trash it!"

Fortunately for Bob, on the climb to the cave I was getting short of breath. Patiently he explained his procedures. His excavations were both modest and controlled. His trenches sampled only part of the cave floor, leaving a large volume of unexcavated fill for future investigators armed with new techniques and new questions. His field team methodically excavated and screened cave earth from 40-inch squares in two controlled trenches. All archaeological materials as well as bones of fish, reptiles, birds, and small mammals, along with seeds and other plant remains, were saved and bagged. In addition, his field crew sieved and saved dung pellets in a systematic fashion. Numerous rocks and boulders and abundant fossil driftwood made their work more difficult. They sampled the cave fill at 2-inch intervals from the surface down to 10 inches, and then at 10-inch intervals to the bedrock below. Pellets from deeper levels were often broken, and there were so many that Bob and his crew did not think they needed to save them all. I subsequently estimated that the cave contained roughly a million artiodactyl pellets.

Poor Bob. There he was, searching for precious artifacts, and what he was finding was mostly fossil goat shit. On top of that, when he invited a devotee of such stuff to see what he had found, he was ordered to save *all* of it. He might well have suggested that I find my own grant, helicopter, and field team. But whatever Bob, an ex–Marine officer, may have thought along those lines, he kept to himself. I cooled down slowly, perhaps influenced by the mellow rose-tinted illumination reflected into the cave from sun on the Redwall across the river. In the end, my colleagues and I were to gather much valuable data from Stanton's Cave, both on this trip and on a one-week return arranged by Bob in September 1970, of which I was delighted to be a part.

I wanted to know what animal had produced the large pellets. We concluded it was indeed Harrington's mountain goat, *Oreamnos harringtoni*, a fossil species related to the living mountain goat, *Oreamnos americanus*. We based this conclusion largely on the bones associated with the pellets. According to Dick Harington (1984), extinct goats (and pos-

Plate 9. Cranium of male Harrington's mountain goat, approximately 13,000 years old, found in a small cave downstream from Stanton's Cave, May 1984. The same specimen is seen in place in plate 10. Photo by Emilee Mead.

sibly living mountain goats) accounted for the majority of the 120 ungulate bones identified from the cave. (The other bones included those of bighorn sheep, bison, and two small species of *Equus*.) The most common artiodactyl bones and horn sheaths found at Rampart Cave were those of Harrington's goats. No other artiodactyls in the fossil fauna would have produced these pellets, which were larger than those of mountain sheep and smaller than those of elk. Since the time of Kellogg, those excavating Grand Canyon caves have commonly found these distinctive pellets associated with bones or horn sheaths of Harrington's goat. In the 1980s, Steve Emslie, Jim Mead, and Larry Coats explored the eastern Grand Canyon and found additional caves with numerous pellets of this size. Future investigators should be able to check this interpretation by DNA analysis.

Our Rampart Cave collaborators, Dick Hansen and Richard Clark, disagreed with us on identification of the pellets. They noted that small elk (wapiti, *Cervus elaphus*) voided similar pellets. Or they might belong to a poorly known extinct cervid, *Navahoceros*, whose bones are found occasionally in caves in western North America. No cervid bones, however, were found in Stanton's Cave. Indeed, I am unaware of any securely identified elk bones from any Grand Canyon cave (Szuter 1991). In any

Plate 10. Floor of cave downstream from Stanton's Cave, natural bone–
midden–goat dung, May 1984. Photo by Emilee Mead.

case, living cervids rarely enter caves, and most of the caves in the Redwall would be inaccessible to them. Other than humans, the only large
animals with climbing skills that might be equal to the challenge were
mountain sheep and mountain goats. (Later in this chapter we will consider the mystery of how the bison bones got there.) I believe that the
large fossil pellets can be assigned with confidence to the extinct goat.
The smaller pellets, less than 0.25 grams in dry weight, might have been
voided by immature *Oreamnos harringtoni,* by mountain sheep (genus
Ovis), or by Rocky Mountain goats, which void much smaller pellets
than did Harrington's goat. Mitochondrial DNA analysis should be able
to resolve this question. But in the absence of their bones, I do not believe that *any* of the pellets found in Grand Canyon caves are attributable to cervids.

When did Harrington's goat last occupy Stanton's Cave? In Rampart
and adjacent Muav Cave, Austin Long and I had estimated the time of
extinction of ground sloths by dating samples from the top of their dung
deposit (see table 5). With Bob Euler's help we could attempt the same
thing here, separating out the large pellets to date the latest occurrence
of the animals producing them. In the stratified pellet profiles, the large
pellets last occur at about 8 to 12 inches below the surface, where they

give way to smaller pellets, presumably of mountain sheep. The youngest sample of the presumed mountain goats was radiocarbon dated at 10,870 ± 200 (see lab catalogue number A-1155 in table 6). A sample of small pellets from the same level yielded a similar result, 10,760 ± 200 (A-1154). The youngest dates on goat pellets and horn sheaths at other sites also approximated 11,000 radiocarbon years ago (Mead, Martin, and others 1986). The youngest measurement, from Crescendo Cave in the Grand Canyon, yielded a date of 10,950 ± 70 (Emslie, Mead, and Coats 1995). Those numbers have a familiar look. The youngest pellets and horn sheaths of extinct goats (see table 6) are similar in age not only to each other but also to the youngest Shasta ground sloth dung at Rampart Cave (see table 5) (Long and Martin 1974).

Most of Jim Mead's data (Mead, Martin, and others 1986) were derived from horn sheaths, keratinous organic material akin to fingernails and toenails, ideal for an uncontaminated radiocarbon measurement. Mead's results ranged from 11,000 to 30,000 radiocarbon years ago, with at least one measurement in every millennium. Many samples comprised dung pellets (10 samples) or horn sheaths (24 samples) found on the surface of cave floors, mainly at Stanton's and Rampart. With rare exceptions, however, his samples could not be selected for youthfulness. He had to take them as they came, and a variety of ages was the result. In contrast, in the buried deposits of Stanton's Cave, we could target the last occurrence of large pellets using classic stratigraphy: the uppermost would be the youngest. The results were in accord with our measurements from Rampart Cave, where we had been able to concentrate on the top of a sloth dung deposit.

The concordance of dates on goat and ground sloth extinction (Mead, Martin, and others 1986) pointed to a common cause, a unique event around 11,000 radiocarbon or 13,000 calendar years ago. As will be discussed further in chapter 8, that was the time of the first well-documented and rapid appearance of prehistoric people, the mammoth hunters (C. V. Haynes 1993; Taylor et al 1996).

As with the ground sloths, we also obtained data on the extinct goat's diet. A Desert Lab student, Frances Bartos King, found no fossil pollen in the pellets, though fossil pollen was abundant in the cave earth. This suggests that the goats occupied the cave in winter, when no plants were in flower. In the summer, they probably migrated to higher elevations. It appears that, like the ground sloths at Rampart Cave, the goats did not occupy Stanton's Cave throughout the year.

Cuticle analysis of the plant remains in the pellets indicated that the

TABLE 6 *Radiocarbon Dates from Stanton's Cave, Arizona*

Lab no.	Date measured	Location and comments
A-1165	2,450 ± 80	5–10 cm; grid I-I
Euler and Olson,	4,095 ± 100	Split–twig figurine
1965		
A-1166	5,760 ± 200	15–20 cm; grid I-I
A-1154	10,760 ± 200	Small fecal pellets (<.25 g);
		20–25 cm; grid I-I
A-1155	10,870 ± 200	Large fecal pellets (>.5 g);
		20–25 cm; grid I-I
A-1167	12,980 ± 200	25–30 cm; grid I-I
A-1082	13,070 ± 470	20–25 cm; grid A-A
A-1132	13,770 ± 500	Large fecal pellets;
		20–25 cm; grid G-G
A-1238	15,230 ± 240	*Teratornis* bone; not *in situ*
A-1168	15,500 ± 600	35–40 cm; grid I-I
A-1246	17,300 ± 800	55–60 cm; grid I-I
A-1056	>35,000	Driftwood at base
		of section below 65 cm;
		grid I-I, east trench

SOURCE: See Robins, Martin, and Long pp. 117–30 in Euler, ed., 1984.

extinct goats ate a variety of shrubs, forbs, and grasses (Robbins, Martin, and Long 1984). Data from other sites confirm this. According to Jim Mead, Mary Kay O'Rourke, and Terri Foppe (1986), grasses, including *Sporobolus, Festuca, Orizopsis,* and *Agropyron,* were a major part of the diet of the extinct goats, along with *Ceanothus* (buck-brush) at certain times. On occasion the extinct goats ate Douglas fir *(Pseudotsuga)* (Mead and others 1987). Harrington's goats at Rampart Cave ate a good deal of juniper (Clark 1977). Rocky Mountain goats dine in the subalpine zone of the northern Rockies, browsing on mountain conifers such as spruce (Peek 2000).

At Stanton's Cave, as at Rampart, packrat middens yielded abundant juniper and other plant remains. Juniper trees must have ranged much closer to both caves than they do at present. The closest junipers I found to Stanton's Cave were far beyond packrat-foraging range. Judging by the fossil pollen record as well, the dominant woody plants outside Stanton's Cave over 10,000 radiocarbon years ago were probably sagebrush, shadscale, and juniper (Robbins, Martin, and Long 1984). The packrat middens indicate some reduction in juniper and sagebrush in the inner gorge of the Grand Canyon around 11,000 radiocarbon years ago, when the goat went extinct. However, a vast area of juniper and sagebrush persists in Arizona and Utah today. As far as one can determine from the remarkable midden records, the habitat and food plants of Harrington's goat persist, even if the animals do not.

Stanton's Cave also harbored other fossil evidence of the late-Quaternary change in climate: a quantity of ancient driftwood. In earlier times an archaeological team might well have tossed it out. Fortunately, as we have seen, Bob took a wide view, and he invited Wes Ferguson of Arizona's Tree Ring Laboratory to examine the wood. Wes (Ferguson 1984) identified it as mainly Douglas fir and cottonwood *(Populus)*. Radiocarbon dating put the age of some samples at more than 40,000 years (Hereford 1984, 105). Cottonwood presently grows wherever it finds slack water along the Colorado River. Thus its occurrence in the more than 40,000-year-old driftwood provides little ecological insight. On the other hand, Douglas fir is presently found upstream in Utah at elevations of 7,000 to 8,500 feet. In Stanton's Cave, the abundance of Douglas fir, combined with the absence of pinyon (according to Ferguson; personal correspondence), the main conifer presently found in beach wrack along the Colorado in the Grand Canyon, at least until the construction of Lake Powell, indicates cooler times when Douglas fir rather than pinyon grew closer to the cave and would have been a major component of its driftwood. Still, it must have taken an extraordinary event, a discharge 33 times the largest historic flood of the Colorado River, to deposit driftwood in Stanton's Cave (Richard Cooley in Hereford 1984, 102).

Stanton's and other caves in the Grand Canyon yielded important records bearing on the mysterious megafaunal extinctions. Grand Canyon caves contained bones not only of large mammals, but also of two giant avian scavengers of those mammals: condors and the even larger teratorns. Stanton's Cave held more than 68 bones of the California Condor *(Gymnogyps californianus,* mainly the large extinct taxon, *Gymnogyps californianus amplus).* Perhaps as many as eight individual condors are represented (see Rea and Hargrave 1984). Elsewhere in the Grand Canyon, a cave exploration team found not only more bones, but also the remains of a fossil nest of a Quaternary condor, revealing what the young birds were fed.

Early investigators believed that the condor fossils they discovered in the Southwest were quite young. Based on their superficial position, in some cases seemingly associated with Anasazi artifacts roughly 1,000 years old, and their fresh appearance ("not petrified"; Phillips, Marshall, and Monson 1964), ornithologists concluded that condors had inhabited the Southwest, including the Grand Canyon, not long ago, during the last millennium or two. This conclusion appeared to be in accord with seven sight records in Arizona and southern Utah reported from the late nineteenth and early twentieth centuries. In March of 1881, for example, a condor was reported from Pierce's Ferry on the Colorado River just west of the Grand Canyon. Alan Phillips and other ornithologists assumed that the birds reported historically were the last representatives of a late-prehistoric population.

Logical as this reasoning seemed, in the end it was to prove once more the importance of radiocarbon dating. Dates on the youngest condor fossils in the canyon were around 11,000 radiocarbon years old. With three exceptions, their dates coincide with those obtained from remains of extinct mammoths, horses, bison, and Harrington's goats (Emslie 1987), and with the association of Clovis points with mammoths in southern Arizona. In 1984, in a remarkable search involving cliff-climbing into otherwise inaccessible caves, Emslie, Mead, and Coats located condor remains in eight such caves in the inner gorge. The youngest in a suite of 18 dates on fossils from New Mexico and West Texas as well as the Grand Canyon was 9,580 ± 160 (AA-790, Emslie 1987). On the surface of a cave much smaller than Stanton's, Emslie found a condor skull so well preserved that it looked fresh (see plate 11), its keratinous beak still attached and its soft parts as hard and dry as beef jerky. An accelerator sample on the soft parts showed the skull to be 12,540 ± 790 (AA-692, Emslie 1987; Martin 1999, 272). The oldest date obtained by Emslie was over 22,000 radiocarbon years. On a preserved condor feather, Larry Coats subsequently obtained a date of over 42,500 years (Coats 1997). California Condors nested in the Grand Canyon during the late Pleistocene, ending at least 9,000 years ago, with no irrefutable evidence (radiocarbon dates on condor remains) of their nesting since, until reintroduced birds began nesting in the twenty-first century.

The dates suggest that after extinction of the mammalian megafauna in the Southwest, condors did not linger very long. In addition to the Pacific Coast, with its rich supply of dead cetaceans, pinnipeds, and salmon, perhaps they persisted in regions that supported large herds of bison. However, there is every reason to suspect that with the disappearance of mountain goats, equids, mammoths, and bison from the Colorado Plateau, the largest avian scavengers such as condors and teratorns lost much of their food supply. Apparently mule deer were the main surviv-

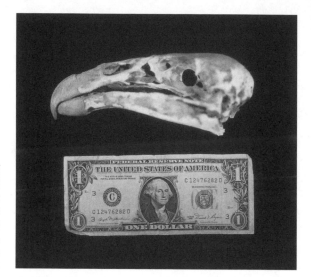

Plate 11. Skull of 13,000-year-old California Condor from small cave downstream from Stanton's Cave, ca. 1985. Photo by the author.

ing large herbivores in the region. There may not have been enough mule deer and rabbit carrion to support any avian scavengers larger than Turkey Vultures.

If the fate of condors was tied to that of large animals, here was more evidence for the extinction of North American megafauna close to or possibly 1,000 years later than that magic number of 11,000 radiocarbon years ago. The three younger dates on condors reported by Emslie (1987) could reflect lingering opportunities for scavengers feeding on remains of bison or lesser beasts that survived the extinctions.

Another of Steve Emslie's 1984 discoveries further supports this conclusion regarding condor food supply. In a packrat midden in the richest cave, Sandblast, Emslie found bones of at least five condors. Bone porosity indicated that some birds were fledglings still in the nest. Food scraps, feathers, and eggshell fragments also indicated nesting. Near a nest were bone fragments of extinct animals: bison, camel, horse, mammoth, and Harrington's goat. According to Emslie, "The bones of large mammals associated with condor remains possibly represent food bones brought to the cave by the condors. Bone fragments of similar size and skeletal elements including phalanges, carpals, tarsals, teeth, and mandibles of horses, cows, sheep and deer, are found in nests of *G. californianus* today" (1987). Despite his caution, there is good reason to believe that Emslie had made an extraordinary discovery of what condors fed to their young before extinction of the late-Quaternary megafauna. How else would such bones have found their way to a condor nest in an inaccessible cave? Bone fragments of cattle, *Bos taurus*, are the most common item in modern condor nests (Collins, Snyder, and Emslie 2000). Bones and teeth supply calcium phosphate to meet the dietary needs of the young.

How, then, to explain the historic sightings of condors in the Southwest, at least 8,000 years after any fossil record of their presence? In the absence of any specimens, the sight records are unsubstantiated. If they are valid they might represent occasional vagrant condors from the Pacific Coast. Condors may also have spread briefly eastward from California as a result of European settlement. Diseases accompanying European contact (Diamond 1997) devastated Native American populations, thus potentially relaxing predation on wildlife and increasing the number of carcasses, to the benefit of predators and scavengers, including condors. Later, the development of Spanish ranching in California in the 1700s and the overstocking of an unfenced range by early Anglo ranchers in the Southwest in the 1880s would also have yielded an ample food supply for condors and other scavengers, more ample than any since the late Quaternary. Under these circumstances condors might have increased in number and perhaps spread eastward, accounting for historic sight records. Nevertheless, in the absence of specimens, these may be questioned. In the twentieth century, fencing and improved range management reduced livestock mortality and thereby diminished resources for scavengers, including condors.

As for the teratorn, that was a discovery I was privileged to witness. In September 1970, toward the back of Stanton's Cave, beyond the zone of artiodactyl pellets and thus beyond that part of the cave with enough light to be frequented by goats and mountain sheep, graduate students Martha Ames and Barney Burns helped me sample cone-shaped fossil packrat middens. The middens were in an unusual spot, in the middle of the cave floor. Perhaps the rats felt safe in the dark from their visually dependent predators and relied on their scent-marked pathways to return to their nests.

One of the middens incorporated a large bird bone. The Los Angeles County Museum identified it as a humerus (wing bone) of the giant scavenger-predator *Teratornis merriami*. Not known previously from Arizona, the species had twice the mass of living condors, with a wingspan (including primary feathers) of about 12 feet, 2.5 feet more than that of a condor (Mawby 1967).

Fossil pollen in matrix scraped from the teratorn bone proved to be 33 percent sagebrush *(Artemisia)*, a higher count than that found in

Holocene cave earth from Stanton's (Robbins, Martin, and Long 1984) or in the modern pollen rain in the inner gorge of the Grand Canyon (King 1973; Mead, O'Rourke, and Foppe 1986). In other words, on the basis of pollen analysis we could expect a pre-Holocene radiocarbon date. Accelerators had yet to be developed, and a large sample, in this case the entire humerus, was needed to provide enough carbon for age determination. After preparation of a plaster cast copy, the humerus was combusted and radiocarbon dated at 15,230 ± 240 (A-1238). This date fell within the range of radiocarbon measurements on other extinct animals, including condors. If condors had difficulties in opening the body cavity of mammoths, bison, or other large carcasses, perhaps the teratorns helped solve the problem. Ecologist David Burney tells me he has seen Lappet-faced Vultures in East Africa rip open carcasses, to the benefit of less powerful scavenging birds.

Quaternary-age fossils of equids, including extinct species, are also found in caves in the West (Harris 1985), though less commonly than fossils of Shasta ground sloths, Harrington's goats, or condors. They include not only bones but also keratinous hooves the size of those of burros—and, provocatively, they first turned up at a time when living wild burros were scheduled to be eliminated from the Grand Canyon (see chapter 10). Fossil equid hooves are known from several caves in Arizona, including Rampart, Stanton's, and Sandblast; from Gypsum Cave and the Eleana Range in Nye County, Nevada; and from sites farther north.

Radiocarbon dates on well-preserved horse metapodials (foot bones) from Alaska range from 20,000 to 12,000 years. Mitochondrial DNA analysis has identified these fossils as *Equus caballus*, the same species at present found in Eurasia and wild in the lower 48 states (Vilà and others 2001). Direct dates on two hooves associated with the ground sloth dung deposit in Gypsum Cave yielded the following results: large (quarter horse–size) hoof (A-1271), 25,000 ± 1,300; small (burro-size) hoof (A-1441), 13,310 ± 210. A date on an *Equus* hoof associated with a stratified sequence of packrat middens from the Eleana Range is 11,210 ± 400, presumably close to the time of equid extinction (Spaulding 1990). Two horse bones collected by Emslie from the fossil condor nest at Sandblast Cave are very likely of the same age as the immature condor bones, 9,580 ± 160 to 13,110 ± 680 (Emslie 1987). According to Emslie, they were among the food scraps brought to the young by their parents. In at least one case, a pair of condors now nesting in Grand Canyon National Park reoccupied a site with evidence of prior, I would suggest late-Pleistocene, occupation.

By 1982 I had almost given up hope of finding my dream: an unstudied cave rich in paleoecological treasure. Then the sun god, Ra, finally delivered on the prayers I had sent up at Bat Cave in 1958. Zoologist Steve Carothers and his friend Loren Haury (the son of Professor Emil Haury, who excavated the first Clovis sites in Arizona), along with National Park Service rangers Larry Belli and Charles Berg, made a monumental discovery. Searching for wild cattle in the Glen Canyon National Recreation Area in southern Utah, they came upon a relatively undisturbed cave, a huge dry rock shelter like a cathedral, with one side open to the sky. Inside, in the spoil of a pit dug by pothunters, were the trampled remains of a bolus that clearly had been of unusual size. Its coarse texture did not resemble that of cow or horse manure. Loren Haury thought it might be dry dung of a ground sloth. Maybe, he thought, it would interest those weird devotees of extinct animal manure at the Desert Lab, who still had not stopped complaining about the sloth shit lost six years before in the Rampart Cave fire.

For weeks a plastic bag containing the precious sample bounced around, along with other field trip leftovers, in the back of Steve's truck. Eventually Steve passed the bag on to Art Phillips at the Museum of Northern Arizona in Flagstaff. Art sent it down to the Desert Lab with a note to the effect that it did not look right for dung of the Shasta ground sloth. It contained what appeared to be masticated segments of coarse grasses, which (except for tall aquatics such as *Phragmites*) the ground sloth rarely ate. Also, the grass stems were up to 5 centimeters (2 inches) in length, longer than the clipped, short plant stems seen in ground sloth dung. The latter, like horse dung, is uniform in texture, and only 50 to 60 percent, rather than 80 to 90 percent, of the fragments are over a centimeter in length (Mead, Agenbroad, and others 1986).

None of the group of Quaternary ecologists at the Desert Lab at the time, including Julio Betancourt, Owen Davis, Pat Fall, Emilee Mead, Jim Mead, Mary Kay O'Rourke, Bob Thompson, Ray Turner, Tom Van Devender, and Bob Webb, could claim to have seen anything like it before. None of the specimens in our sizable reference collection of scats, fossil or modern, resembled the mystery dung sample, with one striking exception: dung samples from African elephants *(Loxodonta africana)* that I had collected in Tsavo Park in Kenya in 1965.

In February 1983 Jim Mead and Larry Agenbroad, a geology professor and proboscidean specialist at the University of Northern Arizona, received National Park Service approval to visit the cave, evaluate its contents, and, if warranted, dig a small test pit. They were guided by Larry

Plate 12. Mouth of Bechan Cave with riparian vegetation in foreground, March 1983. Photo by the author.

Belli and accompanied by Utah State Archaeologist Dave Madsen and Utah State Paleontologist Dave Gillette. They returned bursting with enthusiasm. From a test pit near the north wall they had extracted two large dung balls. One, designated M-1, measured 230 by 170 by 85 millimeters (9 by 7 by 3 inches); the other, M-2, was 225 by 175 by 80 millimeters.

According to Joe Dudley (1999), mature African elephant bulls produce the largest dung balls of any elephant, 190–230 millimeters (7 to 9 inches) in diameter. The dung balls from the locality to be known as Bechan Cave (derived from a Navajo word for "big feces") were somewhat flattened from trampling or from the weight of overburden. But their size greatly exceeded that of our Shasta ground sloth samples and approximated those of samples from female African and Asian elephants. Also unlike our ground sloth samples, the Bechan Cave boluses were not segmented, were not encased in a dried mucosoid coating, and with rough handling threatened to fall apart. We began to suspect that this was the spoor of America's second-largest extinct mammal—the Columbian mammoth. (The largest, the imperial mammoth, is very rare or absent from the fossil record in Arizona.)

The first two radiocarbon measurements on the dung yielded the following results: M-1 (A-3212), 11,670 ± 300, with a delta carbon 13 of 23.2 percent; M-2 (A-3213), 12,900 ± 160 with the same delta carbon

13, 23.2 percent (Davis and others 1984). Although the dates were close to being significantly different, the carbon 13 values of the dung balls were identical. Pollen samples from the two, analyzed by Owen Davis, were also similar, within expected statistical error. The pollen and delta carbon 13 similarities, plus the fact that both dung balls came out of the same small test pit, led me to suspect that both were dropped at the same time by the same individual.

In the spring of 1983 Owen and I returned with Larry and Jim to help them plot an isopatch (thickness of deposit) map based on the contents of 49 auger holes. The probes disclosed a buried organic layer, mainly mammoth dung, up to 16 inches thick (Agenbroad and Mead 1996) and estimated to contain 14,000 cubic feet of dung, more than the deposit at Rampart Cave before the fire.

Fourteen accelerator radiocarbon dates on boluses collected that spring range from 11,870 ± 140 to 12,880 ± 140 and average 12,450 years. It is possible that part of the dung blanket was deposited rapidly, perhaps in no more than a few days, by a small matriarchal herd seeking shade in the hot season. Larry Agenbroad and Jim Mead (1996) accept the maximum and minimum dates as valid, which would indicate occupation of Bechan Cave by mammoths at least sporadically over 1,800 years, ending 11,600 years ago. In either case, the data suggest abandonment of the cave shortly before 11,000 radiocarbon years ago, unless we failed to notice and date a younger layer. In Bechan Cave the last boluses deposited were not as obvious as in the case of the surface dung balls at Rampart, a much smaller collecting surface to sample. I am not confident that we can determine the time when mammoths last entered this cave.

Beyond one tooth of a medium-sized bovid, probably the shrub ox *Euceratherium,* we found no bones of extinct megafauna at Bechan Cave. We did find coarse hairs matching those from woolly mammoths in Siberia and Alaska. While it seems certain that all the dung we examined came from the Columbian mammoth, there is a remote possibility that some other megaherbivore, perhaps a medium-large ground sloth, such as *Paramylodon,* was responsible for part of the deposit. A Shasta ground sloth could have been the source of a small dung ball containing an acorn.

Nearly one-third of the plant macrofossils Owen Davis found in the dung samples were sedge *(Carex)* seeds (achenes) from marshy habitats or standing water. These were followed in abundance by cactus spines (17 percent)—the first evidence, to my knowledge, that American mammoths ate cactus—and grass florets (12 percent). Wood fragments included birch *(Betula,* 12 percent), rose *(Rosa,* 11 percent), saltbush *(Atriplex,* 5 percent), sagebrush *(Artemisia,* 3.5 percent), and smaller amounts of blue spruce *(Picea pungens),* snowberry *(Symphoricarpos),* and red osier dogwood *(Cornus stolonifera).* The fossil deposit even yielded the spruce cone gall, *Chermes cooleyi* (illustrated in Davis and others 1984). Associations of some of the plants indicated in the dung samples, including blue spruce and water birch, are found today along streams in the Henry and other mountains in southern Utah at elevations of 7,300 to 8,000 feet, 3,000 feet higher than Bechan Cave. Some of the plant fossils associated with the Bechan Cave dung are also extralocal. Evidently, the deposit accumulated at a time when the climate was cooler than at present and trees or shrubs such as blue spruce, water birch, and red osier dogwood grew in riparian habitat outside the cave.

Not all the twigs and sticks from the dung unit necessarily came from the mammoths. Several other kinds of animals, especially the large packrat *Neotoma cinerea,* very likely introduced twigs and branches. Nevertheless, African elephants can switch from grazing to browsing according to season and habitat. Possibly some of the woody material in the dung blanket represents the digestive residue of browsing Columbian mammoths.

To obtain a better estimate of the mammoths' diet, Owen Davis dissected 25 fragments of boluses under 7x magnification (Davis and others 1985). The identifiable plant remains were removed from the matrix and weighed. Over 95 percent of the boluses constituted a graminoid (grassy) matrix composed of crushed culms (stems) and leaves of grasses, sedges, and rushes *(Juncus),* along with small amounts of sand. The remainder was dominated (88 percent) by saltbush wood and fruits, followed by sedge achenes (5 percent), cactus parts (4 percent), and wood of sagebrush (1 percent).

The presence of saltbush, cactus, and sagebrush indicates dry upland vegetation when the dung layer was being deposited. Pollen analysis supports the interpretation that upland vegetation at the time was sagebrush steppe with blue spruce and water birch along the drainages (Agenbroad and Mead 1996; Davis and others 1984). At present the upland supports xerophytic shrubs, especially blackbrush. Wetlands harbor cottonwood, willow, sedges, and other aquatic herbs. In Davis's words, "The abundance of aquatic plants [especially sedges and rushes] in the dung demonstrates the importance of the riparian community to the diet of the mammoths. . . . Riparian vegetation near Bechan Cave may have attracted mammoths to the site. At the time Paleoindians reached southern Utah, mammoths and other megafauna may have been concentrated along streams and other mesic sites in an otherwise arid landscape" (Davis and others 1985).

Bechan Cave also offered an opportunity to test another analytical technique. In large numbers, herbivores may leave a trace in the fossil record not only of pollen, but also of spores. Ponds surrounded historically by heavily grazed land capture runoff rich in manure. This is the substrate for a distinctive fungal spore type known as *Sporormiella*, a small, smooth-walled spore in the shape of a pistol bullet with a sigmoid aperture. Davis and Pete Mehringer first reported *Sporormiella* from Wildcat Lake in eastern Washington, near a historic sheep pasture. Then Davis found it in fossil deposits of 16,000 to 12,000 radiocarbon years ago, just predating the megafaunal extinctions. At Bechan Cave he recovered large numbers of *Sporormiella* spores from the mammoth dung boluses themselves: 2,390 spores per cubic centimeter, equivalent to 16 percent of the pollen count (Davis 1987). Robinson (2003) is finding *Sporormiella* in lake muds associated with fossil bones of mastodon and stag moose *(Cervalces)*. After mastodon extinction the spores decline, as Davis (1987) anticipated.

It turns out that all of the Colorado River drainage was mammoth country, from low elevations in southwestern Arizona near the Sea of Cortez to high elevations (9,000 feet) at the headwaters of the Colorado River in northern Utah. Larry and Jim found fragments of mammoth dung in four other Utah rock shelters. The youngest samples dated at 9,000 to 11,000 radiocarbon years, the oldest at 26,140 ± 670 and 28,290 ± 2,100 years (Mead and Agenbroad 1992). (Because mammoth dung is spongy, not compact, there are a variety of ways it might be contaminated by younger organic material, which would account for dates of less than 11,000 radiocarbon years.) Meanwhile, Dave Madsen and Dave Gillette had jurisdiction over the Huntington Canyon site, near the crest of the Wasatch Plateau in central Utah. At 9,000 feet, this is to my knowledge the highest elevation at which extinct megafauna have been found anywhere in the United States. Reliable radiocarbon dates on bone organics of 11,200 and 10,800 years ago dated an old mammoth, suffering arthritis and fused vertebrae, and a giant short-faced bear.

Spectacular as the Bechan Cave finds were, it was Dick Hansen (1980) who had first identified fossil dung of a mammoth. In the summer of 1975 Jesse Jennings led a University of Utah field school at Cowboy Cave in Wayne County, Utah. Located at 5,800 feet in a short, nameless box canyon near Canyonlands National Park, the cave harbored a stratified cultural deposit up to 5 feet thick. The deposit proved to be rich in perishable artifacts such as sandals, basketry, medicine bags, and prehistoric shelled corn, suggesting seed stock. A test pit also revealed chopped plant remains in a culturally sterile unit at the base. Radiocarbon dating proved these to be at least 3,000 years older than the oldest cultural remains in the cave. Five dates on the material ranged from 11,020 ± 180 (A-1660) to 13,040 ± 440 (A-1654) years. Although in reverse stratigraphic order, the values are conformable with the end of the reign of the late-glacial megafauna. Geof Spaulding and Ken Petersen (1980) suspected a clerical error was responsible for the reversal. The portion of the deposit associated with the dated material yielded both pollen and macrofossil evidence of spruce and Douglas fir, trees now found only at higher elevations.

Most of the identifiable residue, rich in finely chewed and digested grasses, looked like cow manure and was attributed to bison. Jennings sent samples of the dung to Dick Hansen's lab. It reported 73 percent dropseed (a grass in the genus *Sporobolus)* and 12 percent sedge. Modern bison eat almost exclusively grasses and sedges (Shaw and Meagher 2000). Three large boluses that Hansen believed to represent mammoth yielded even more dropseed, 95 percent. Two samples identified tentatively as horse also yielded 95 percent dropseed, suggesting that instead of horse they too might be mammoth dung. A broken piece of tusk roughly 4.5 inches in length, recovered from the base of the dung blanket, supported Hansen's identification of mammoth dung. Hansen also suspected the presence of Pleistocene *Equus.* The samples in storage at the University of Utah deserve DNA testing for the kinds of extinct animals once present in Cowboy Cave.

Many more fossil sites in Arizona and adjacent states feature mammoth remains. Others provide support for the overkill theory by linking the mammoths to the earliest hunters in America, the Clovis people. Nevertheless, some archaeologists have claimed that there should be more than are known. As I learned in the 1950s in Canada, many archaeologists search for the oldest. Recently some have proposed that the First Americans crossed the Atlantic in relatively quiet water between icebergs of glacial age. Others propose migration down the west coast of North America to South America. The traditional view of entry from Siberia through Beringia and into Alaska, with eventual passage through an ice-free corridor during deglaciation, is considered passé in some circles, whose members favor a coastal entry past melting glaciers on the Pacific coast of southern Alaska and Canada.

The heart of the argument for me is that late-Quaternary climatic change, while impressive, is essentially no different from what we see in many, many swings from cold-dry to warm-wet and dusty to dust-free climates in the last 700,000 years or so. Unless oceanographers, ice-core stratigraphers, and climatologists find some unique event, the classic approach to explaining Quaternary extinctions by some physical means is inoperable.

Human involvement in the extinction process also encounters objections. Is the chronology tightly timed to the spread of Clovis hunters? Why are there not more kill sites? We will return to these issues in the latter half of this effort at following John Alroy's call.

SIX

DEADLY SYNCOPATION

In regard to the wildness of birds towards man, there is no
way of accounting for it except as an inherited habit . . .
both at the Galapagos and at the Falklands, [many individu-
als] have been pursued and injured by man, but yet have not
learned a salutary dread of him. We may infer from these
facts what havoc the introduction of any new beast of prey
must cause in a country before the instincts of the indigenous
inhabitants have become adapted to the stranger's craft
or power.

**Charles Darwin, *Journal of Researches . . .
During the Voyage of the* Beagle**

As we have seen, the basis for the overkill model is what Ross MacPhee
calls the "deadly syncopation" of human arrivals and megafaunal extinctions in new lands. Before we explore some of the arguments raised against
this model, it will be useful to review that syncopation in more detail.

Although geochemical dates are not available for the times of extinction of all the target species (those whose disappearance is of interest relative to human arrival), the trend is quite clear. The fossil record
shows no concentrated extinction of large mammals until we reach near
time (Alroy 1999). Then, within the last 50,000 years—where we have
the advantage of a much more refined time scale, thanks in large part
to radiocarbon dating—megafaunal extinctions pop up independently
in different parts of the world (Martin and Steadman 1999). From a
modest start in Africa and Eurasia one to two million years ago (reaching Flores by one million years ago; Morwood and others 1998), they

erupt in near time in the following sequence: Australasia, the Solomon Islands, continental America, the West Indies, and Pacific islands from New Caledonia east to Hawaii and Rapanui (Easter Island). They end (apart from historic losses) in New Zealand and Madagascar, with moa extinction in the former approximately 500 years ago (Bunce and others 2003) and hippopotamus extinction in eastern Madagascar 200 years ago.

Large-animal extinctions on the continents of human origin, Africa and Asia, were relatively few and episodic, not only in near time (see totals at bottom of table 3) but over the last several million years (Klein 1999). In Africa, extinctions of large mammals were last apparent around 1.5 to 2.5 million years ago, during the early evolution of the genus *Homo*. A moderate extinction pulse involving a small number of species blighted Europe and Asia over the last 70,000 years. Some of these were simply extirpations, or local extinctions; close relatives survive elsewhere (Stuart 1999). Mammoths disappeared gradually in Eurasia, over thousands of years (Stuart and others 2002). In less than a millennium they disappeared from North America, leaving a small group of survivors on the Pribilof Islands in the Bering Sea. Even when summed, the Afro-Asian losses extending over more than a million years do not match the number of near-time extinctions in either Australia, the Americas, Madagascar, or New Zealand.

The large number of large mammals in Africa, many more than in America, has long been accepted as a basic fact of zoogeography, a natural condition and a baseline. Historically, no genus of large mammal in the New World features more than a few species, and no part of the Americas could begin to match the game plains of Africa for numbers of large mammals, especially artiodactyls. Above all, the New World lacks anything to match Africa's elephants, black and white rhinoceroses, hippopotamuses, and giraffes, five familiar species of living megaherbivores.

Rarely considered, at least until recently, is the fact that in the late Quaternary North America north of Mexico had at least 10 species of megaherbivores (five proboscideans, two ground sloths, two glyptodonts, and a camelid). This is twice Africa's quota. Beyond that, South America was more than twice as rich as North America, with an extraordinary assemblage of 25 species of megaherbivores, those exceeding 1,000 kilograms (2,200 pounds). These include four species of proboscideans, nine of ground sloths, five glyptodonts, five notoungulates (all but one in the genus *Toxodon*), and two camelids (Lyons, Smith, and Brown 2004). In addition, South America was the evolutionary center for phorusrhacoid birds, highly specialized predators with a beak like that of a hawk or an eagle but much larger (Murray and Vickers-Rich 2004).

For its part the Afro-Asian fauna is much richer than that of the Americas, either in the present or in near time, in medium-sized herbivore species, those ranging from about 0.5 to 45 kilograms (1 to 100 pounds) in body mass (see figure 1 in Lyons, Smith, and Brown 2004). For example, Africa has 18 species of duiker (genus *Cephalophus*), small antelopes ranging from the mass of a hare to that of a small female elk. Similarly, 16 species of gazelle (species *Gazella*) weighing between 12 and 85 kilograms (25 to 185 pounds) are found in Africa and parts of Asia. Nothing like them exists among the larger (over 45 kilograms) or small mammals in the Americas. In comparison, the most species-rich genus of small artiodactyls in the New World is the brocket deer *(Mazama)*, with four (possibly six) species. Though the current distribution of mammalian fauna has long been accepted as simply the way things are, it has been greatly influenced by different extinction rates on different continents over the last two million years and especially in near time.

Under the overkill model, two arguments may explain the lower extinction rate on the continents of human origin. In the Old World hominids evolved with other mammals, which accordingly developed both fear of human predators and other defenses against them. As our hominid ancestors moved into drier or colder and often less hospitable areas, such as the high plateaus of central Asia and the coast of the Arctic Ocean, contact between large mammals and Paleolithic people was limited.* Taxa of *Rangifer* (caribou or reindeer), *Ovis* (mountain sheep), *Bison* (bison), *Cervus* (red deer), and *Alces* (moose, elk) had a lengthy exposure to Paleolithic hunters and may have evolved more resistance to them than any of the genera of large mammals suddenly encountering Stone Age humans entering the Americas, especially South America, where genetic influence from Old World fauna was negligible. It is no surprise that South American large mammals were obliterated.

Data from Australia, the first continent to be invaded by humans, are crucial to the overkill model. In the 1960s the first radiocarbon dates appeared showing the potential time of extinction of Australia's megafauna. The youngest dates on diprotodonts were 7,000 and 13,000 years ago.

---

*In Africa and Asia, human diseases such as sleeping sickness and tropical malarias may also have favored the survival of large mammals. Elephants, rhinoceroses, and hippos survived in more tropical areas but not in temperate, boreal, and/or subarctic parts of Eurasia, where human pathogens are less numerous or less virulent.

The dates were overshots and not replicated by later geochronology. It was the same sort of problem I encountered in trying to assemble a chronology of megafaunal extinction in North America over 40 years ago. The first dates released were seriously inaccurate, a common failing of bone samples not properly treated.

Recently Australian geochemists and paleontologists released a series of thermoluminescence dates on their extinct fauna (Roberts and others 2001). The method lacks the precision of radiocarbon. Nevertheless, none of the dates were younger than 46,000 years. While generally accepted, an age of 46,000 for contact between humans and extinct fauna is challenged in one stratified site at Cuddie Springs in New South Wales (Wroe and others 2004). But the vast majority of Australian evidence indicates that large mammals, large birds, and large reptiles became extinct around 42,000 to 48,000 years ago (Gillespie 2002; Roberts and others 2001). And *Homo sapiens* reached Australia roughly coincident with these extinctions, as we will see in chapter 7 (Mulvaney and Kamminga 1999). This discovery contradicted those who once assumed, decades ago, that not until the early or mid-Holocene, just a few thousand years ago, would ancestral Aborigines have been clever enough to make the boats needed to cross the ocean from Southeast Asia to Australia.

In the Americas, the time of extinction of about half the lost genera remains to be determined critically. However, there is no chronological indication that any of the extinct American genera listed in tables 2 and 3 endured after 13,000 calendar years ago. These genera are absent from numerous fossil deposits (including bone beds) of the last 10,000 radiocarbon years excavated in North, Central, and South America. In addition, the youngest radiocarbon dates on these genera from many localities terminate at around 11,000 radiocarbon years ago. The boundary is a sharp one. Pending further geochronological testing, the extinction of ground sloths and Harrington's mountain goats appears to be particularly well dated.

From other extinct animals in caves in the Southwest, Ken Cole, Donna Howell, Jim Mead, Geof Spaulding, Bob Thompson, Tom Van Devender, and other Desert Lab researchers have also recovered dates not appreciably younger than 11,000 radiocarbon years ago. As we have seen, the youngest reliable dates from both Argentina and Chile are quite similar (see table 5).

Back in the 1950s and 1960s I flirted with much younger (and much older) radiocarbon dates on extinct North American megafauna. Many cultural associations linked artifacts with bones of extinct mammals. Archaeologists reported human artifacts or at least cultural charcoal mixed with the bones of extinct North American animals radiocarbon dating from 2,000 to 30,000 years ago (Martin 1958b), a result that proved erroneous at both ends. There were bugs to work out, especially with charcoal, which was not always in a true association with the animal bone investigators dearly hoped to date. The problem lingers. In youthful abandon I did what armchair consumers of published radiocarbon date lists often did in those early days. We had not visited, much less excavated, the fossil sites or dealt with any of the fossil and stratigraphic evidence. We had no basis on which to presume to pick and choose. If professional archaeologists or geologists found associations of artifacts with remains of extinct species, I accepted their dates.

Taken literally, those dates indicated that people and extinct animals had coexisted for thousands of years. From *Science,* for example, I extracted a Florida date on charcoal with the lab catalogue number Lamont 211 that was said to be associated with extinct mammals. If accurate, this would be extraordinary: extinct large animals still alive in Florida only 2,000 years ago (Martin 1958b)!* As far as I know, L-211 still lies unmourned in the sizable geochronological graveyard of anomalous, undefended, or unreplicated radiocarbon dates. In the 1950s, more dates suggested that mammoths and other extinct megafauna lived as late as 8,000 radiocarbon years ago. Libby's C-222 date of 8,500 years on Shasta ground sloths at Gypsum Cave was an example. But this date could not be replicated (Poinar and others 1998). And with new suites of radiocarbon dates, such as the series on ground sloths, extinct mountain goats, and condors from the Grand Canyon and elsewhere in the Southwest, and the steady input of new dates around 11,000 radiocarbon years on extinct horses, camels, saber-toothed cats, and mastodons, the chronology supporting extinctions 8,000 or fewer years ago crumbled (Martin 1990; Stuart 1991).

Much additional chronological work is needed in the West Indies,

---

*An outraged Florida archaeologist of impeccable credentials, R. P. Bullen, concluded, based on his experience excavating many archaeological sites, that extinct mammals could not possibly have survived in Florida until only 2,000 years ago. No good associations of cultural remains with extinct animals that late, or even thousands of years older, were known to exist in Florida. The problem with L-211 might well have been intrusion of charcoal from a younger hearth into a much older deposit of extinct animal bones. Though in most cases charcoal is a reliable source of carbon for radiocarbon dating, an association problem of this type could lead to misinterpretations. Whatever the explanation, no defenders of L-211 emerged, and ten years later, without comment beyond citing Bullen's article, I washed my hands of it (Martin and Wright 1967).

which, like mainland North and South America, seem to lack robust associations between cultural material and bones of extinct species. At least some of the dwarf ground sloths there lasted until roughly 5,000 years ago, and archaeologists place the first human arrival in the Caribbean Islands at about 6,000 years ago (Wilson in Fagan 1996). The fossiliferous asphalt seep in Matanzas Province in Cuba includes four genera of dwarf ground sloth (Iturralde-Vinent and others 2000). It also includes *Ornimegalonyx*, a nearly flightless giant raptor or "walking owl" up to 18 kilograms (40 pounds) in weight—possibly one of Cuba's top terrestrial carnivores, second in size only to the terrestrial crocodiles. It includes, as well, an extinct condor, *Gymnogyps varonai*—perhaps Cuba's largest extinct scavenger—and an extinct crane, *Grus cubensis*. Ross MacPhee supplied samples of a dwarf ground sloth, *Parocnus brownii*, from the seep for radiocarbon dating. The results ranged from 4,960 ± 280 years at the youngest to 11,880 ± 420 years at the oldest. At 4,960 to 6,330, MacPhee's youngest dates are "apparently in good agreement with the trend of the earliest archaeological dates for this island" (MacPhee, Flemming, and Lunde 1999).

After the West Indies, extinctions next erupted in Madagascar. Recent chronological refinement suggests that Madagascar's giant birds, hippos, giant tortoises, and giant extinct lemurs vanished around the time of human arrival from Borneo over 2,000 years ago. Recent work by David Burney (Burney and others 2004) indicates extinctions within the last 2,400 years in the megaherbivores on Madagascar's west coast. Presumably extinction of hippo and the elephant bird, *Aepyornis*, accounts for a reduction in the dung fungus *Sporormiella* and a rise in charcoal. The discovery of the sensitivity of this spore type to the presence of dung and dung production by large animals may prove to be the most valuable tool in the methodology of those seeking to detect and date both extinctions and change in biomass.

Around the same time, as we have seen, extinctions struck remote islands in the Pacific, followed by New Zealand less than 1,000 years ago. For the relatively few islands undiscovered by prehistoric voyagers, such as the Galapagos in the southeastern Pacific, the Commander Islands in the north Pacific, the Mascarenes in the Indian Ocean, and the Azores in the Atlantic, evidence of prehistoric extinctions is minimal or unknown. In some cases there is evidence that the same species went extinct on different landmasses at different times, coincident with human arrival. For instance, in New Caledonia, the last records of the horned turtle, *Meiolania*, are associated with cultural remains about 1,500 years old; this genus had gone extinct tens of thousands of years earlier in Australia (Martin and Steadman 1999, 27–28).

Recent discoveries of island extinctions are particularly interesting. Almost without notice, beginning in the mid-1970s, avian paleontologists of the Smithsonian Museum reported bones of new taxa of an extinct goose-sized flightless duck and a flightless ibis from Hawaii (Olson and Wetmore 1976). Since then, thanks especially to paleo-ornithologist David Steadman and his archaeological collaborators, a completely unexpected flood of fossil finds, especially the bones of extinct birds, has turned up on islands and archipelagoes in the remote Pacific. Investigators from Darwin onward (summarized in Quammen 1996) had simply missed finding the fossils. Therefore, only recently has it been possible to evaluate these rich faunas within an archaeological framework (Kirch and Hunt 1997). As it turns out, humans played an unexpectedly traumatic role in a severe prehistoric reduction of the fauna of Pacific islands (Burney and others 2001; Kirch and Hunt 1997; Steadman 1995, n.d.).

Endemic Pacific island birds included taxa of parrots, pigeons, doves, megapodes or bush turkeys, and especially flightless rails. Thousands of these taxa, as well as many seabird colonies, are no more. Over 2,000 taxa of flightless rails *(Rallidae)* alone have been lost (Steadman 1995, 1997, n.d.). The 800 Micronesian, Melanesian, and Polynesian islands over a square kilometer in area have together lost roughly 8,000 species, taxa, or indigenous populations of land birds (Martin and Steadman 1999, 29). Other terrestrial vertebrates and numerous taxa of endemic land snails have also disappeared. Islands in many other places as well have lost small endemic mammals, including some that had evolved from larger continental mammals. Many of these endemics disappeared in the Holocene, typically when brought into contact with continental species of mice or rats *(Rattus)*.

Pacific island fossils occur in cave deposits or open sites on all islands inhabited prehistorically. When they have been radiocarbon dated, most of the extinctions fall in the last few thousand years, and all of these correlate with the arrival either of humans or of the Pacific rats that accompanied the voyagers (MacPhee and Marx 1997), often within a few hundred years (Martin and Steadman 1999, 29). From west to east across the Pacific, beginning 3,000 years ago in New Caledonia and Tonga, passing through Hawaii and Rapanui (Easter Island) 1,500 years ago, and ending only 700 years ago in New Zealand, extinction marked the spread of our species.

In fact, directly or indirectly, human colonization resulted in more near-time vertebrate extinctions on Pacific islands than on the continents (Steadman 1995, n.d.).* According to Dave Steadman, "On any remote Pacific island with a respectable fossil record, one can expect to find at least two to three times the number of land bird taxa before contact than after" (Martin and Steadman 1999, 29). The evidence of syncopation on particular islands is quite clear. For instance, there were far fewer if any near-time extinctions on the Hawaiian islands before colonization than after (Martin 1990). In New Zealand, radiocarbon dating suggests that "human hunting and habitat destruction drove the 11 species of moa to extinction in less than 100 years after Polynesian settlement" (Holdaway and Jacomb 2000). (The moa fauna has recently been reduced from 11 to 10 species by recovery of nuclear DNA sequences, and further reduction is expected [Huynen and others 2003].) All 14 New Zealand birds weighing over 9 kilograms (20 pounds) disappeared, including moas up to 180 kilograms (400 pounds), along with about 28 of the 140 species of birds under 9 kilograms (A. Anderson 1997; Worthy and Holdaway 2002).

Exceptions to the general pattern, not only in the Pacific but also in the other oceans of the world, are instructive. Islands lacking artifacts or other evidence of prehistoric inhabitants (including severe prehistoric extinctions) include the Azores and Bermuda in the Atlantic, the Mascarenes in the Indian Ocean, Lord Howe Island east of Australia, the Commander Islands in the Bering Strait, and the Galapagos west of South America. Such islands serve as controls for an anthropogenic extinction model.

In addition there are "islands of doom," occupied and abandoned prehistorically. These islands, including Norfolk, Henderson, and Pitcairn in the South Pacific and Nihue in the Hawaiian chain, had been colonized and then, apparently after exhaustion of their resources, abandoned. The most dramatic example of resource depletion was on Rapanui (Easter Island), where the wood for canoes was depleted and surviving settlers engaged in internal warfare. This suggests prehistoric humans' capacity to exhaust resources in a touch-and-go.

Worldwide, the body size of animals scales to the size of the landmass under consideration: the largest animals on smaller landmasses are smaller than those on large landmasses. This is also true of the body size

---

*MacArthur-Wilson island biogeography (developed by eminent ecologists Robert MacArthur and E. O. Wilson) employs mathematical models to explain why certain islands supported richer faunas than others. I suggest that we recognize "Olson-Wetmore island biogeography" after Storrs Olson and Alexander Wetmore, the first to show massive extinctions on oceanic islands accompanying human arrival.

of *extinct* animals. For example, in North and South America the preponderance of extinct mammals exceeded 45 kilograms (100 pounds), while Madagascar, the size of Texas, lost mammals down to 9 kilograms (20 pounds). The West Indies lost many more medium to small mammals than either continental North America or Madagascar, and oceanic islands, especially in the remote Pacific, lost animals down to the size of land snails.

This scaling is consistent with the overkill model. If we make the reasonable assumption that human hunters took the path of least resistance, they would first have gone after those species that were easiest to track, find, and kill and that provided the most food (or prestige) for the least effort. In general, these would have been the largest animals in any given region. Human foragers would have worked their way down the size scale until they reached species that were difficult to kill. In addition, it would have been easier to find smaller animals in smaller areas. Finally, any other ecological disruption caused by humans (e.g., an increase in fires) would have had greater impact in smaller areas. For all these reasons, the size scaling of the extinctions supports the idea of overkill.

In short, the global pattern of extinctions in near time appears to be just what one might expect if people played the major role in triggering them. If this "deadly syncopation" was a coincidence, if the extinctions had nothing to do with our species and its global hegira, their "true" cause will be the greatest geological discovery of the new millennium.

The syncopation we have seen, however, would be consistent not only with overkill but with "overill": the idea that the extinctions were caused not only by human hunting but also by other debilitating changes introduced by humans, such as new commensals (Steadman 1995) or pathogens (MacPhee and Marx 1997). In some cases it seems clear that introduced rats caused extinctions.

A dramatic case has emerged from New Zealand. Some 55 species of flightless or ground-nesting birds vanished after Pacific rats *(Rattus exulans)* appeared 2,000 years ago (Worthy and Holdaway 2002). Very likely as stowaways in double-hulled sailing canoes crewed by Polynesian explorers, the rats crossed the vast Pacific, jumping ship on islands free of terrestrial mammals. While not endorsed in all quarters, recently obtained radiocarbon dates on Pacific rat bones indicated their presence in New Zealand before human settlement (Holdaway and others 2002). These rats apparently eliminated the smaller members of New Zealand's extinct fauna, such as flightless wrens and "giant" insects. In addition, presumably because of rat predation, 57 species of ground-nesting petrels, Storm Petrels, and other Procellariiformes (an order of predominantly pelagic birds) abandoned breeding colonies in New Zealand. Most managed to survive by virtue of nesting colonies on rat-free islands elsewhere in the Southern Hemisphere (Steadman 1995, n.d.).

For whatever reason, the Polynesians themselves seem not to have settled New Zealand until at least a millennium after their initial landfall. Then they rapidly eliminated 10 species of moa. Human hunting forced moa extinctions; the loss of smaller species may involve the Polynesian rat, introduced dogs, and other side effects of human colonization, such as increase in wildfire.

Of the various factors that could have contributed to overill, rats probably receive the most scholarly attention. Microorganisms are more controversial, partly because they are harder to detect in the fossil record. Corbett (1973) proposed that prehistoric humans had unwittingly spread viruses akin to Ebola, which can burn explosively through populations of large animals. Similarly, Ross MacPhee and Preston Marx (1997) have proposed hyperdisease, the inadvertent introduction by the first human invaders of highly lethal pathogens able to jump the species barrier. They note that disease appears to be a factor in at least some historic extinctions, as illustrated by the loss of two endemic species of *Rattus* on Christmas Island south of Java. But most of the "extinctionists" I have consulted consider hyperdisease at best an unlikely cause of the worldwide extinctions of large mammals. Its role appears to be testable by mitochondrial DNA analysis, however, and MacPhee has assembled a paleovirology team and is searching for lethal diseases in possible victims, such as the youngest known mammoths, those found on Wrangel Island.

Other indirect anthropogenic causes of extinction have also been proposed. These include Dan Janzen's theory that "the Pleistocene hunters had help," which Charles Kay has refined using Canadian data on wolf-moose predation (Kay 2002). Under this model, as the proboscideans declined, predators such as the giant short-faced bears, scimitar cats, and American lions would have turned to smaller prey (Janzen 1983; Kay 2002). Bob Dewar (1997) has proposed that wild cattle helped force extinctions in Madagascar. And in some cases, the precise reason that a species died off after human colonization is simply not known. For example, the giant rats of the Galapagos went extinct shortly after the Spanish discovered the islands. There are no rat kill sites, but some sort of anthropogenic linkage, rat-borne diseases included, is strongly suspected.

Overkill does not, of course, purport to explain every extinction since the evolution of *Homo sapiens*. Although firmly established examples are few, some late-Quaternary extinctions apparently occurred before human arrival on the landmasses in question and thus must be ascribed to other causes. To be certain of this, robust chronologies and, ideally, DNA evidence are needed.

In my initial brief treatment of the West Indies, for example, I assumed that all known extinctions corresponded with the arrival of prehistoric people. More critical appraisal reduced the number of such cases. For instance, according to MacPhee and others (1989), extinction of the giant rodents *Amblyrhiza* in Anguilla and *Clidomys* in Jamaica, which I assumed reflected overkill (Martin 1984), predated human arrival. If Manuel Iturralde-Vinent and Ross MacPhee (1999) are correct that most land mammal lineages entered the Greater Antilles around the time of the Eocene-Oligocene transition, or at any time in the Tertiary, a large number of Tertiary turnovers (extinctions) predating near time, and hence human arrival, can be expected. In the majority of cases, however, the extinctions of endemic West Indian mammals are not chronologically separated from possible human presence and from prehistoric human activities. Undoubtedly, significant extinctions accompanying evolution occurred on all the large and persistent island platforms rising out of deep water. Such change is not incorporated in the overkill model and may be difficult to detect if the fossil record is poor, as is often the case for insular faunas that predate near time. Overall, insular faunas of extinct vertebrates provide a valuable opportunity to test anthropogenic and other extinction models.

SEVEN

DIGGING FOR THE
FIRST PEOPLE IN AMERICA

High Stakes at Tule Springs

> New theories appear, new arguments rage, and a fully
> satisfying solution has not been reached. The peopling
> yarn is still coming in installments, like the Pickwick
> Papers, without a plot or denouement.
>
> **Gary Haynes, *The Early Settlement of North America:***
> ***The Clovis Era***

The "deadly syncopation" argument does not work, of course, if the dates of near-time extinctions do not correlate with the dates humans arrived in new lands. Particularly in the Americas, the latter are the subject of ongoing debate. The prevalent opinion is that foragers or hunters suddenly appeared in the western United States around 13,000 years ago; their archaeological sites can be dated by geochronological techniques (C. V. Haynes 1991, 1993). They left artifacts, including Clovis spear points, distinctive fluted blades tightly lashed to wooden spears or throwing sticks and used for thrusting and throwing. As I discuss in chapter 8, Clovis points are occasionally associated with mammoth bones.

Some anthropologists, however, believe that people reached America well before 13,000 years ago. These claims have generated tremendous interest among prehistorians and in the media. If people did arrive and grow to appreciable numbers thousands of years before the extinction of the megafauna, fewer objections could be made for a noncultural explanation for the extinctions.

Through the early 1960s, I accepted the claims of some archaeologists of an early human occupation of the Americas. Based on various publications, especially the first publications of radiocarbon dates, over ten millennia appeared to separate the initial human arrivals in the New World from the last occurrences of extinct megafauna. The former arrived 20,000 years ago or much earlier; the latter vanished by 8,000 years ago or even later. I had no reason to doubt the claims of professional archaeologists I had met and in some cases interviewed, such as Alan L. Bryan, Ruth Gruhn, George Carter, Tom Lee, Scotty MacNeish, and Ruth Simpson. I listened to their talks and read their reports of archaeological sites that long predated Clovis time. Then some intensive efforts to confirm this early occupation of the Americas yielded negative results and I turned into a skeptic.

In the late 1950s, Ruth Simpson, an archaeologist with the Southwest Museum in Los Angeles, undertook excavations of colluvial fans (accumulated rock detritus and soil) in the Calico Hills, in the Mojave Desert east of Los Angeles. Interpreting sharp-edged cherts (selected from a very large quantity of fan gravels) as artifacts, she reported finding an ancient pre-Clovis site. Other archaeologists were dubious.

Then the famous African archaeologist and paleontologist Louis Leakey inspected Simpson's excavations and endorsed her finds (Leakey, Simpson, and Clements 1968). Simpson invited a blue-ribbon panel of professional archaeologists and geomorphologists to inspect the site. They did, and gave it a thumbs-down. Although closely resembling manmade tools (see Tankersley 2002, 192, photograph of a pseudo-artifact produced by earthquake-generated liquefactions at the Calico Hills site), the cherts were widely scattered, rather than concentrated, as one would expect of artifacts in an archaeological site. In addition, they could not be distinguished from rocks in transport in a fan, which may fracture naturally. The archaeologists were also bothered by the burial of artifacts supposedly 40,000 years old in alluvial fans that geologists viewed as at least 400,000 years old. Although some people still regard Calico Hills as an ancient cultural deposit, most professional archaeologists dismiss it, including some who accept certain other claims of pre-Clovis colonization. This is only one example of the many proposals of an early New World antiquity that have failed to pass the test of tangible and reproducible evidence (Martin 1974).

An earlier test of the pre-Clovis invasion hypothesis had failed. The circumstances seemed so promising at first: the site, Tule Springs, would nail down the presence of people and extinct animals long before 11,000 radiocarbon years ago. In those years before explosive urban growth, Tule Springs lay just northwest of the tourist mecca of Las Vegas (it has since been built over by it). In 1933 the Southwest Museum conducted excavations there, turning up an obsidian flake of human manufacture in association with remains of an extinct camel. Mark Harrington, the excavator of Gypsum Cave in the 1920s, was part of this team. At Tule Springs Harrington found the bones of extinct bison and other species associated with charcoal, presumably cultural in origin, and on this basis reported early man. Then, in the early 1960s, Willard Libby began searching for groundbreaking new applications of his radiocarbon dating method. Two earlier radiocarbon dates, one by Libby, the other by Wally Broecker and Larry Kulp of the Lamont-Doherty Earth Observatory at Columbia University (see Wormington and Ellis 1967, 3) suggested that humans were present at Tule Springs over 23,000 years ago. These dates cried out for verification. Libby convened an ad hoc committee of leading archaeologists and geologists of the region; at their suggestion, the Southwest Museum returned to the site for multidisciplinary excavations between October 1, 1962, and January 31, 1963 (Wormington and Ellis 1967).

The Tule Springs project was funded in part by the National Science Foundation, along with massive assistance from the private sector. Herschel Smith, a Southern California contractor with an active interest in archaeology and geology, offered the services of the construction industry gratis. International Harvester provided two of the largest bulldozers then manufactured in the United States. Allis-Chalmers contributed a large motor scraper. Pafford and Associates of Los Angeles made available aerial photos, surveyed a grid, and prepared a detailed contour map. Union Oil donated all fuels and lubricants. Members of the International Operating Engineers Union, Local 12, ran the heavy equipment, donating their time (C. V. Haynes 1967b, 16).

Several of my friends and colleagues participated in the Tule Springs dig, and I was as excited about it as they were. Dick Shutler, by then with the Nevada State Museum, helped organize the project and became its field director. He invited Vance Haynes and Pete Mehringer, at the time grad students in the geochronology program at the University of Arizona, to join him. Both were enthusiastic, talented, and experienced. Funds were available to bring in leading archaeologists of the time to inspect the results.

With all that support, Dick, Vance, and their team opened a remarkable 3-mile maze of trenches, in some places to a depth of 30 feet through strata well over 15,000 radiocarbon years old. Vance had the glorious opportunity to map the cuts, fills, and soils and (with some 80 radiocarbon dates) determine in detail the chronology of the exposures. Even now such an abundance of controlled radiocarbon dates at such a crucial site is exceptional. Forty years ago it was state of the art. In addition, the field team could count on weekend turnaround of radiocarbon dates from Libby's lab. Normally such service takes weeks or months. This control on stratigraphy allowed at least some of the uncertainties in correlating units to be resolved as the units were being uncovered and mapped.

With abundant radiometric and stratigraphic control, Pete Mehringer extracted and analyzed fossil pollen from the exposed alluvium and from sectioned spring mounds. The mounds form when dust storms intercept artesian springs. The conifer pollen and plant macrofossils gave evidence of biotic and climatic change during the Quaternary. Meanwhile, the paleontologists identified fossils of mammoths, large camels, large and small horses, ground sloths, pronghorn antelope, the extinct American lion, and a teratorn (Mawby 1967).

All these were valuable data well worth obtaining. But when Vance Haynes analyzed the organic material that Harrington had considered to be cultural charcoal, it proved to be naturally oxidized plant material, soluble in alkaline washes. It was not charcoal. Equally disappointing, no evidence of chipping of lithics or workshop activity could be detected at the site. There were no Clovis points or, for that matter, spear points or knives of any kind. The entire recovery of artifacts was minuscule. According to Haynes, "All I know of are a single scraper from Locality 4 and a crude caliche bead (perhaps natural) and polished bone awl tip from Locality 3" (see Wormington and Ellis 1967, 39, 360). Moreover, nothing to match or replicate the date of more than 23,800 years ago (lab catalogue number C-914) for the extinct bison discovered by Harrington, allegedly with cultural remains, could be verified, and the stratigraphic overlap between the oldest artifacts and the youngest extinct faunas narrowed to a unit dated at 13,000 to 15,000 calendar years ago.

The members of the Tule Springs team and their consultants had hoped that the magnificent new sections would yield direct evidence of hunting or butchering of extinct animals. The National Geographic Society was ready to trumpet such a find; in anticipation of success, one of the society's best artists, Jay Matternes, prepared a dramatic color illustration of a kicking camel being speared by early hunters. (The illustration ended up as a black-and-white frontispiece to Wormington and Ellis 1967, a classic monograph that is still in print.) Despite the great expectations and three months of careful field work by a superbly supplied team overseen by the best professional archaeologists in the West, the excavations at Tule Springs satisfied none of those dreaming of an early archaeological site, particularly a kill site. But they did yield something more fundamental: an appreciation of negative evidence. It is true that "the absence of evidence is not evidence of absence"—but there are limits to how long and how strongly one can keep believing when supporting evidence is lacking. Apparently, professional archaeologists can not all agree on what constitutes an archaeological site, a disagreement that erupts from time to time at archaeological sites throughout the Americas, in recent years no less than in the days of Tule Springs.

Now a new set of sites and a new generation of advocates champion pre-Clovis inhabitants of the New World. The "sites" are the result of an ardent search over the last 20 to 30 years, on the heels of conspicuous failures at Gypsum Cave, Calico Hills, Tule Springs, and elsewhere, along with the embarrassment of Sandia points (see p. 146). Among some of those who advocate for early sites and trumpet a "paradigm shift" in the absence of solid evidence, I detect a passion similar to that exhibited by cryptozoologists in their eternal search for Bigfoot. I acknowledge that the subject of pre-Clovis colonization is one for experienced archaeologists and geologists to resolve. However, this time around, unless and until the heady claims can be replicated independently by skilled skeptics, I am going to stay on the sidelines.

From time to time colleagues ask why I do not accept Tom Dillehay's Monte Verde site in Chile, or Jim Adovasio's Meadowcroft site in Pennsylvania, as predating Clovis (see Bonnichsen and Turnmire 1999). In the view of Adovasio and David Pedler (1997), another main contender is Bluefish Caves, Yukon; there are perhaps a dozen more (Lavallée 2000; Roosevelt, Douglas, and Brown 2002), including Pedra Furada in Brazil. But none has been excavated and verified independently by neutral (or skeptical) parties. This is no reflection on the optimism of the original investigators; no one should be denied the opportunity to search for the unknown and to report their discoveries. But the fact remains that no claim for pre-Clovis archaeology has been put to what I call the "Tule Springs test." Sadly, many sites have been totally excavated by their claimants. Scotty MacNeish deserves credit for responding positively to my urging that he not completely excavate Pendejo Cave east of Oro Grande, New Mexico, on the Fort Bliss military reservation in southern New Mexico, a site he pronounced to be over 30,000 years old. I was amazed at how few professional archaeologists came to have a look at Scotty's claim, in particular those who are convinced that such evidence exists. Although he was a lifelong professional archaeologist trained at the University of Chicago and one-time president of the Society for American Archaeology, many members of his profession interested in early sites did not bother to inspect his site. (For an insightful, experienced overview of pre-Clovis claims, I recommend chapter 1 in G. Haynes 2002b.)

In the past decade geologists Jay Quade, Vance Haynes, and Erv Taylor accepted Scotty's kind invitation to visit the site and were less than convinced. The problem was not a matter of its antiquity—radiocarbon dates and extinct fauna are among the evidence for a pre-Clovis age—but of the claim for human occupation. I hope that other archaeologists will also recognize their responsibility to spare significant portions of their sites, ideally at least half, for independent, reasonably unbiased verification teams. Such have not been permitted to excavate Meadowcroft in Pennsylvania. The famous Monte Verde site in Chile was visited by a team of outside archaeologists only after excavation terminated. Not all in the group agreed on the claims of antiquity. Moreover, if there was a pre-Clovis population in Chile 13,000 radiocarbon years ago, in Pennsylvania 18,000 radiocarbon years ago, or in Alaska even earlier, as has been claimed, those brave pioneers did not demonstrate the environmental adaptations seen at the end of the Stone Age (Upper Paleolithic) in the Old World. Where are the large numbers of large sites with many distinctive stone tools? Where are the cave drawings? Where are the Upper Paleolithic huts made of concentrations of mammoth bones (see Klein 1999, 538–540)? We know Clovis hunters of mammoth occupied North America 11,000 radiocarbon years ago. Had people been here at an earlier time and lived as their Upper Paleolithic ancestors did in Asia, there should be no difficulty finding archaeological sites older than Clovis.

The heart of the argument appears to be that while the chronology of megafaunal extinction falls in the late glacial and in well-dated deposits close to 11,000 radiocarbon years ago, the age of Clovis points, there are few kill sites, most of them of mammoths. Archaeologists impressed with the claims for much older archaeology in the New World wash their hands of the matter of megafaunal extinction by assigning it to climatic change, rarely if ever the special research interest of those making this interpretation. Can this dilemma be resolved?

The contrast of the New World with the Australian experience is striking. In 1962 John Mulvaney discovered 16,000-year-old cultural charcoal in Kenniff Cave in central Queensland. Within four years, radio-carbon dates associated with cultural material for Koonalda Cave, Burrill Lake on the New South Wales coast, and three rock shelters in Kakadu National Park all demonstrated occupation older than 20,000 years.

The 1970s brought still earlier dates for human occupation on the shore of Lake Mungo in western New South Wales and at Devil's Lair, a cave in southwest Australia. These findings established firmly a prehistory of more than 30,000 radiocarbon years and the possibility of at least 40,000. By 1980 over 20 sites of similar vintage had been identified. By 1999 Mulvaney and Johan Kamminga recognized more than 150 Pleistocene-age archaeological sites (Mulvaney and Kamminga 1999, 136). On the hypothesis that a flush of charcoal from vast and intense brush fires would occur during or very soon after human invasion, Peter Kershaw and others (2002) reported such evidence in marine sediments dating back about 42,000 years, supporting the archaeological record and another consequence of colonization, the known time when Australian megafauna suffered mass extinction.

By 1980 new discoveries were running into the limits of radiocarbon dating as it could be applied at the time. At Lake Mungo human skeletons had been found and dated to over 30,000 radiocarbon years. According to Richard Gillespie (n.d.), they are now thought to exceed 40,000 years. The Australian record has yielded early sites from Tasmania in the south to the vicinity of Darwin in the north and from Perth in the southwest to Cookstown in the northeast. Extinction of the Australian megafauna, once thought to have occurred later in time, is dated in the 40,000-to-50,000-year bracket. If there is a lengthy gap between the time of human arrival in Australia and prehistoric extinctions, robust evidence for such a chronology has yet to appear. In addition, the Australian extinctions took place tens of thousands of years earlier than any known extinctions of megafauna (moas) in New Zealand, hippos, elephant birds, and giant tortoises in Madagascar (Burney, Robinson, and Burney 2003), or ground sloths in South America. Therefore, whatever caused extinction of megafauna in Australia, the Younger Dryas or any other worldwide climatic pulse of the last 20,000 years can be ruled out, since it would have affected these areas as well.

Australia is about the size of the contiguous United States. Its population is more than an order of magnitude smaller, which translates roughly into about a tenth the number of archaeologists searching for early sites. Australia's primary productivity is lower, meaning its mean annual production of plant dry matter is less than in the United States.

As a result, there would have been smaller numbers of prehistoric people to leave fossils or artifacts. Finally, there has been no vast program of salvage archaeology to expose buried sites that might yield new discoveries. Despite these handicaps, Australian archaeologists have in the last four decades radically extended their chronology of human arrival to or beyond the limit of radiocarbon dating at dozens of sites, while archaeologists hot on the trail of pre-Clovis colonization have failed to nail down any robust evidence of North American sites that is acceptable to the community of archaeologists as a whole. The discrepancy should trigger serious revisionary thinking. Perhaps American archaeologists in search of pre-Clovis sites need to hire some Australians. Aussies seem to be capable of finding and agreeing on the existence of sites tens of thousands of years older than the late-glacial fluted points and fishtail points that are the oldest artifacts unclouded by controversy in the Americas (G. Haynes 2002b).

But there is more to my position than the lack of pre-Clovis artifacts. I have grave doubts about the existence of a widespread and biologically effective human population in the Americas before 13,000 years ago precisely *because* large, slow-moving, eminently huntable animals such as ground sloths continued to occupy their favorite dung caves in North and South America as late as they did. The youngest dates on ground sloth dung may have as much to say about human presence as the oldest dates on artifacts. I admit that such inductive reasoning takes me onto treacherous ground. For example, Alejandro Garcia (2003) recently reported radiocarbon dates from La Gruta del Indio in Argentina that are 2,000 years younger than accepted dates on ground sloth extinction in America. Did these huntable ground sloths escape discovery and destruction by the first people into southern Argentina? A 9,000-year date on survival of ground sloths in North or South America is almost as noteworthy as the discovery of pre-Clovis archaeology would be.

Those of us not only skeptical of claims for human arrivals long before Clovis time in the late glacial but also willing to entertain the idea of human involvement in the extinction process find ourselves labeled as conservatives by European archaeologists who happily accept much older claims and choose not to consider the possibility of overkill (Lavallée 2000). Nevertheless, I think that what we know and can deduce of the behavior of various species in the late Stone Age, including our own, supports my argument. Once a species as adaptable as ours entered a continent as rich in resources as America, what would have prevented our species from immediately exerting an extraordinary impact on large, vulnerable native animals, in particular the ground sloths? An understanding of our species and its capabilities, as well as the morphology of giant xenarthrans, reinforces the theoretical case, as I see it, for overkill hard on the heels of first human arrival.

Any hypothetical colonists reaching the New World well in advance of Clovis time would have had to be inept indeed to leave no ecological trace of their presence. Gifted with the natural resources of a continent of Eden, a land extraordinarily rich in edible wild plants and animals, would the first humans fail to multiply and adapt? This would indeed be puzzling, since their economies must have been derived from those of the late-Paleolithic hunters and foragers of Eurasia. The stone and bone tools of those peoples are commonly found in association with bones of large mammals. Bison, reindeer, and red deer were popular prey, as evidenced by large numbers of Old World sites (Klein 1999). In contrast, our hypothetical pre-Clovis "flower people," as Danièle Lavallée's "radicals" appear to view them, must have ignored the doomed large animals in the New World. Instead, we are led to believe that the self-styled radicals imagine the First Americans tapped the resources of the Americas so modestly that humankind managed to remain scarce for thousands or tens of thousands of years before the large mammals suddenly vanished (to the great surprise, no doubt, of these early innocents).

Even if these First Americans had nothing to do with the extinctions, there is every reason to expect they would somehow have been involved with large animals, such as by using ivory or bones in crafting art or constructing huts (C. V. Haynes 1991). But little if any evidence of the use of animal remains has been detected to support the concept that humans populated America appreciably before 11,000 radiocarbon years ago. In the Old World, where far fewer animals became extinct, the bones or ivory of large extinct animals are abundantly present in an archaeological context, including in huts of mammoth bones constructed by late-Paleolithic foragers.

And what of the animal-related art? Old World cave paintings and stone and ivory carvings are widespread. They date back to and beyond 40,000 years ago, the limit of radiocarbon dating. In the Aurignacian (about 40,000 years ago, when the first modern Europeans appeared), some cave artists magnificently portrayed woolly mammoths, woolly rhinoceroses, giant deer, and other now extinct or extirpated mammals (Klein 1999). There is none of this in pre-Clovis or Clovis America. Clovis sites with mammoth or mastodon remains are tightly constrained to 13,000 years ago (Taylor, Haynes, and Stuiver 1996). In the New World, archaeological remains are virtually unknown in secure association with extinct animal remains. For example, we found no archaeological remains associated directly with the mammoth dung layer in Bechan Cave, which was a few thousand years older than Clovis time.

Some archaeologists discount the skills, the abilities, the very genius of the Clovis pioneers, depicting them as timid, tentative, and diffident foragers baffled at first by the major changes in climate and vegetation to be found as they spread south through the Americas. I find it much more likely that the first people here were skillful, robust, accomplished, highly adaptable, and above all, persistent and very likely passionate hunters. Their remote ancestors had overrun the forests and savannas of the Asian and African tropics, the shrub lands of the Mediterranean, the mixed conifer and oak forests of central Europe. More recently they had penetrated the cold, wind-swept high plateaus of Central Asia, the boreal taiga, and finally, in late-glacial Beringia, the arctic and subarctic steppe tundra—where winter temperatures *now* may drop to 50 degrees Fahrenheit, 51 degrees Celsius, below freezing—with minimal resources for months at a time. In the process of crossing the Bering Land Bridge into the New World, the early Americans must have found many animals familiar to them, such as woolly mammoths, musk oxen, bison, horses, caribou or reindeer, wapiti, and Dall's sheep, as well as the unfamiliar mastodons and megalonychid ground sloths, both relatively scarce. The American animals had no prior experience of the new invaders and, like elk in Yellowstone National Park, long separated from wolves and no longer fearing them, almost surely responded fearlessly to the sight and scent of these strange bipeds.

Very likely, wolfish dogs and/or ravens *(Corvus corax)* accompanied the First Americans and helped the invaders locate prey (Heinrich 1999). In lower latitudes they could have found fresh kills by observing scavenging birds such as condors, eagles, teratorns, and vultures. They could have followed fresh game trails to watering holes, especially those of proboscideans; droughts would have offered them particularly good opportunities for locating and dispatching mammoths (Jelinek 1967; G. Haynes 2002b). The foraging habits of native herbivores would have shown them some of the edible native plants. In addition, they must have had the geological and geographical insight to locate and remember the best outcrops of cryptocrystalline rocks, such as cherts, widely sought for making stone tools. The early people knew lithology.

In short, the idea of *Homo sapiens* as a subdued or ineffective inhabitant of the Americas, one who had the skills to get here but not the biotic potential to make a difference after arrival, in particular to neglect to hunt desirable and vulnerable large prey, seems to me simply absurd. As they arrived, people immediately became the keystone species.

Beyond the skills of our late–Stone Age predecessors was the unusual vulnerability of many large animals in America and on any other land-mass unknown to hominids. Among the large animals, ground sloths should have been among the most vulnerable to human predation. Being large and leaving large and distinctive droppings, the ground sloths and their relatives, the glyptodonts, would have been easy to locate and track. Moreover, the sloths most likely defended themselves by sitting up and clawing their attackers, as do their modern relatives, the giant ant-eaters. Presumably some such strategy would have served ground sloths for millions of years against contemporary carnivores, but human hunters would soon have learned to stay out of reach of the slow-moving animals and would have speared or stoned them to death from a safe distance.

Even if the sloths shared the giant anteater's other defense of a foul taste, their unusual vulnerability would have tempted younger hunters or their preteen followers to use the luckless animals for target practice. Similarly, the armored carapace of the pampatheres and glyptodonts suggests a passive defense that might ward off big cats and dire wolves, and the glyptodonts' clublike or macelike tails must have packed a lethal wallop against unwary attackers. Neither defense would have lasted long against adaptable human hunters.

Turning to persistent proposals of a pre-Clovis culture, the question is how humans could have been skulking around the hemisphere for thousands of years *without* depleting the megafauna or hastening extinctions. Harrington's goats were better suited than the ground sloths to escape human predators, so why did they succumb? To be sure, mountain goats at the generic level did survive; Harrington's goats, living at lower altitudes, may have been more exposed to the newcomers than Rocky Mountain goats.

Joel Berger's recent findings suggest that at first contact the American fauna would have lacked behavioral defenses against humans, including the fear and alarm response necessary to inspire potential prey to fight or flee. Berger, Swenson, and Persson (2001) have reported a naïve response by moose (normally highly suspicious creatures) and elk when wolves were reintroduced to the Yellowstone region after an absence of at least a century. Very likely it would have taken longer for potential prey to learn to fear the new human predators than it did for the moose to learn to fear the wolves, which their ancestors had known to be dangerous.

Plate 13. Mammoth and Clovis hunters: *The Dance of Death* (mural, now destroyed), by Dr. Wallace Woolfenden.

In general, naïve prey species are utterly unprepared for the intense predation humans can inflict. In the undisturbed Galapagos Islands, Charles Darwin wrote of doves so tame they were killed steadily and left in a growing pile by a boy with a switch who waited for them at a spring (*Voyage of the* Beagle): "The birds are Strangers to Man & think him as innocent as their Countrymen the huge Tortoises" (quoted in Terrell 1986, 94). Sea lions, Darwin's Finches, and many other animals in the Galapagos are still famous for their fearless behavior in the presence of humans. I believe they give a fair indication of how the large animals of any uninhabited lands, islands or continents, might have responded to their first contact with humans (see plate 14).

A similar example can be found in the Gauttier Mountains, an isolated, uninhabited, and largely unvisited range rising out of tropical lowlands in New Guinea. It was the tree kangaroos in these mountains that led ecologist Jared Diamond to shed his doubts about the possibility of prehistoric overkill in America. Diamond wrote:

> Until I had worked in the Gauttiers, I was mystified to understand how the few Maoris in the vastness of New Zealand's South Island could have killed all the moas, and how anyone could take seriously the Mosimann-Martin hypothesis of Clovis hunters eliminating most large mammals from

Plate 14. First contact: The author and sea lion, Santa Fe, Galapagos Islands, 1981. Photo by M. K. O'Rourke.

North and South America in a millennium or so. I no longer find this at all surprising when I recall the large kangaroo *Dendrolagus matschiei* remaining on a tree trunk at a height of 2 meters, watching my field assistant and me as we talked nearby in full sight. The low densities of these mammals elsewhere in New Guinea, even in areas visited annually only by nomadic hunters, illustrate how susceptible large, k-selected mammals with low reproductive rates are to hunting pressure. (Diamond 1984, 847)

Interestingly, many of the North American large mammals to survive human contact originated in or were closely related to species found in the Old World, where they overlapped with humans (Kurtén and Anderson 1980). Examples include caribou *(Rangifer)*, elk *(Cervus)*, moose *(Alces)*, and mountain sheep *(Ovis)*. When they first arrived the moose, elk, caribou, and bison that we now consider natives of North America were every bit as foreign to the continent as the humans who followed soon after. The extinction of large mammals of North America eliminated mostly long-established natives. Most of the newcomers survived (Ward 1997; Kurtén and Anderson 1980).

Some are very reluctant to accept that our species killed off the large mammals. Would the first people have been prudent predators, adjusting their harvest to their needs? Or would they have killed freely, without restraint, if prey did not attempt to escape? We cannot know for certain one way or the other, but we do have some interesting data on how predatory species behave when killing is easy.

From his research on the Serengeti Plains of Tanzania, ethologist Hans Kruuk (1972) records an excessive slaughter of Thompson's gazelle by spotted hyenas. On a moonless and stormy night in November 1966, hyenas killed 82 gazelle and badly injured 27 more in a single 4-square-mile area. Kruuk reconstructed the killings "from tracks in the wet mud, injuries of the victims, and other evidence." Interestingly, Kruuk noted, "Of a sample of 59 dead ones, 13 had been partly eaten (almost only soft parts). . . . Tracks indicated that spotted hyenas had walked very quietly from one victim to the next at a normal walking pace. . . . When gazelle are hunted by hyenas in the usual manner, there is a long and fast chase before the gazelle is caught, over distances of up to 5 km." Presumably the gazelle had been dazed and traumatized by the storm, thus becoming uniquely vulnerable to the hyenas, which had not hesitated to kill far more than they could eat.

Kruuk's account set off a flood of literature on "surplus killing" or "excessive killing." Wildlife ecologists discovered that wolves fed on only half of the carcasses of newborn caribou they killed in minutes in a 1-square-mile area of the Northern Territories of Canada (Miller, Gunn, and Broughton 1985). In deep snow in especially snowy winters in Minnesota, wolves killed more deer than they could consume (Del Giudice 1998). Especially during seasons when the animals were under stress, making their dispatch even easier, might humans have been similarly disposed to surplus killing of their naïve American prey? Some research suggests an affirmative answer: "In recent years anthropologists and conservation biologists have suggested that the hunting strategies of subsistence hunters are opportunistic, not density dependent or designed for sustained yield" (Kay 1994; see also Winterhalder and Lu 1997 and references therein). Both historically and prehistorically, an opportunistic hunting strategy reduced or suppressed high-ranked (highly desired) resources in parts of western North America (Kay 1994; Truett 1996; Broughton 1997; Martin and Szuter 1999).

In addition, anthropologists have noted that what might appear to be wasteful killing may not be. Much of the carcass of a large animal is inedible; consuming too much protein can even be poisonous. People active in an outdoor life have high caloric requirements; often their prime need is fat. In times of drought and for much of the late winter and spring, bison, and presumably other Quaternary mammals, would have been in poor condition, with minimal body fat. Most of the carcass would have been unfit to eat, as Lewis and Clark discovered. Thus, the historic killing of bison for the fat-rich tongues and hump alone may not have been as wasteful, at least in lean seasons, as John Speth (1983) observed. The foot pads of proboscideans would have been another tempting source of fat.

Occasionally the argument may be heard that First Americans were skilled conservationists who would not have exterminated potentially valuable and attractive species of big game. There is little doubt about First Americans' abilities as shapers of habitat; the ethnographic and paleoecological evidence for knowledgeable manipulation of fire, in particular, is overwhelming (Bonnicksen and others 1999; Davis and others 2002). But this does not prove that the First Americans were incapable of surplus killing. Some prefer to think of them as vegetarians who lived in harmony with nature. They would not have done violence to large animals, and certainly would not have exterminated the "gentle giant" ground sloths. Perhaps the elders would have sought to protect ground sloths. Who knows? But there is no reason to believe that the first hunters could perceive, much less control, the negative impacts of their arrival. Enjoying an ample food supply and not threatened by any serious diseases or enemies themselves, the hunters would more likely have rapidly increased their numbers, expanded their range, and eliminated their more vulnerable preferred prey.

Emotional objections to a view of our ancestors as surplus killers may influence some theorists. As Peter Murray has written, "The notion that aboriginal hunters may not have been conservation-minded will be obstinately denied by those committed to the idea. Climatic change is a conveniently neutral causal factor that can extinguish a megafauna without any emotive connotations" (Murray 1991, 1141). Alas, though peaceful coexistence is a condition greatly to be desired, yearnings are not enough to create it.

Let me make clear that identifying Quaternary humans as agents of mass extinction denigrates neither the ancestors of present-day Indian or Aboriginal people nor the rights of present-day Indian nations to manage their game resources as they see fit. I am horrified to be told that the theory of overkill has been used against both Native Americans and Australian Aborigines in managerial controversies. The most that can be said is that our species, *Homo sapiens,* appears to have been involved. We may blame our species, for all the good that will do. Indeed, should we, in the next 12,000 years, cause as few extinctions of large mammals as the Native Americans have in the 12,000 calendar years since the days of the ground sloths, we would be able to consider ourselves incredibly lucky. Alas, the record of American conservation in the years since Columbus suggests we will fall far short of that standard—with a vast number of smaller species at risk even now, as *Science* and other sources report evidence of the onset of Earth's sixth mass extinction.

A final argument sometimes raised regarding the overlap of humans with extinct mammals is that humans could not have spread rapidly enough to account for virtually simultaneous extinctions in North America and southern South America, at least not without leaving field evidence of having done so (Jelinek 1967; Meltzer 1993). Those archaeologists who are gradualists look for equilibrium between resources and human populations. Nothing would be expected to happen catastrophically, and an empty continent would be populated very slowly. Once upon a time I too ruled out the possibility of a catastrophe, both for large mammals in the late Quaternary and for dinosaurs at the end of the Cretaceous (Martin 1967b). However, to me the evidence (in both cases) is now too strong to ignore. The archaeological record simply does not preclude the possibility of a prehistoric blitz in which the invaders swept the hemisphere in 1,000 years or less, leaving dozens of extinct taxa in their wake (particularly larger, more slowly reproducing species) (Martin 1973).

Models of maximum population growth rates in a favorable environment of previously unhunted animals have yielded some fascinating results. In 1973, with help from Dave Adam and other young faculty in the geosciences department at the University of Arizona, I whipped off a back-of-the-envelope article in which I maximized the rate of human invasion and the magnitude of impact of hunters while minimizing the time required to attain a large population and to sweep the continent. I knew some would find the parameters extreme, but the editors and reviewers of *Science,* bless their hearts, accepted the article.

A few years later, Jim Mosimann, a biometrician with the National Institutes of Health and a close friend from graduate school days, designed a more respectable model based on difference equations and the work of Russian climatologist Mikhail Budyko (1967). To our knowledge Budyko was the first to treat mathematically the extinction of mammoths by human predation. Using then state-of-the-art software running on an IBM 650, we generated different versions of a discovery scenario (Mosimann and Martin 1975). These were revised by Stephen Whittington and Bennett Dyke (1984) and most recently by John Alroy. Alroy (2001) found he could attain rapid extinctions of many large North American mammals with a much smaller human population and a more modest kill rate. Graeme Caughley (1988) and Richard Holdaway and

Charles Jacomb (2000) have devised versions for New Zealand, and Steve Mithen (1997) has modeled mammoth extinction in Eurasia.

The computations are actually fairly simple. We can start with a very small group—say, 100—arriving from Asia. The maximum rate of population growth observed today anywhere in the world is roughly 3.3 percent per year, or a doubling every 22 years. Historical records of population growth on newly colonized tropical islands are consistent with this figure. In a large and lush New World, well supplied with resources and relatively free from contagious diseases (most of which appear to have originated in the tropics), an annual growth rate of 3 percent is not unreasonable. Anthropologists have estimated that a hunting population requires, for its support, at least a square mile per person, if the game supply is ample (Ward 1997, 146). This means North America could have supported roughly one million Clovis people. At a 3 percent annual growth rate, the Clovis invaders would have reached this number in only 350 years (20 generations). At a growth rate half as high, they would have taken 800 years.

Geographic expansion would have required far more modest movement at any one time than may at first appear. It is probably safe to assume that the Clovis people moved outward from their camps as easy-to-hunt game became locally scarce. If they moved only 10 miles a year, they would have reached the Gulf of Mexico in 350 years. Progressing in this fashion, they could have caused the mass extinctions without ever even reaching their theoretical population maximum. They would simply have abandoned each region after hunting it out. (A succession of such short-term stays would also help explain why they left so little trace of their presence.) After the megafauna were gone, the human population may have crashed unless people rapidly learned alternative survival skills, such as fishing, hunting smaller game, and gathering.

TRICKS, HOAXES, AND BAD SCIENCE

Some caring people may find the overkill model disquieting on a personal or spiritual level. They may feel it denigrates native people in those parts of the world, such as the Americas, where prehistoric extinctions appear to have been anthropogenic in origin. Being uncertain of the details and being aware that the vast majority of extinctions occurred before the existence of humans on the planet, social scientists may simply decide that environmental explanations must be the answer to the mystery of megafaunal extinctions. Climate change would be a prime example of a non-anthropogenic, environmental explanation.

In an extreme case, Native American anthropologist and activist Vine Deloria dismisses the scientific approach entirely. He treats a potential anthropogenic explanation for megafaunal extinction in America as a direct attack on his people's religious beliefs. Taking a position parallel to the Christian fundamentalists' explanation of the origins of humans, Deloria claims that the ancestors of America's Indians did not cross the Bering Land Bridge and discover a continent. Instead he asserts, "We were always here" (Deloria 1995). He urges that we look to the various Native American creation stories for answers to where the Native Americans' ancestors came from. Deloria rejects the notion that the First Americans exterminated the mammoths and other megafauna. Indeed he rejects what he characterizes as the Eurocentric science that supports the idea, including most of the fossil record and much of archaeology and biogeography. He dismisses radiocarbon dating, telling us that geochemists in radiocarbon labs are white liars who can invent dates.

This book is not the place to argue the relative merits of social scientific and scientific approaches to paleontology. Suffice it to say that I do not find Deloria's approach of any help in unraveling the mysteries of megafaunal extinctions.

Of course, no researcher is entitled to a free ride. Hoaxes, frauds, and counterfeits can plague paleontology (Tankersley 2002) as other disciplines, and when they do, it is important to acknowledge them as such. For example, archaeologist Frank Hibben claimed to have found what he named "Sandia points," allegedly older than Clovis points, at Sandia Cave, New Mexico. For some time Sandia points earned serious treatment by archaeologists, as in *Ancient Man in North America* (Wormington 1957), a popular book that long served as the final word on fluted-point archaeology. However, the dates originally assigned to the Sandia points have since been discredited (Preston 1995). Hoax or not, the existence of Sandia, or any other points older than Folsom and Clovis, has yet to be recognized.

In November 2000, prominent Japanese archaeologist Shinichi Fujimura, known as "God's Hands" because he was so successful at discovering early sites, was revealed to have planted at least some of them himself (G. Haynes 2002b).

The most famous hoax is Piltdown man, a human skull and an orangutan jaw, treated chemically to look like fossils of the same individual, discovered in England in 1911. Geochemistry, including radiocarbon dating, helped uncover the hoax decades later. The identity of the hoaxer, or hoaxers, remains a matter of speculation; discoverer Charles Dawson, famed paleontologist Sir Arthur Keith, and author Sir Arthur Conan Doyle (who lived near the discovery site) are among the many suspects.

As a final example, the University of Arizona radiocarbon laboratory found reason to believe that a sample submitted for dating of rock varnish might have been tampered with. Rock varnish is a veneer on many exposed rocks and boulders, especially in arid lands. Some imagined that it could be used in geochronology and claimed the dates supported pre-Clovis colonization of the Americas. The specimen the laboratory received for dating resembled a mixture of charcoal and coal. When split, the particles that looked like coal proved to be too old to measure by radiocarbon dating, as one would expect of coal. The material that looked like charcoal was modern. Charges of misconduct were contested by the plaintiff's lawyers and eventually settled out of court.

This unusual case underscores the obvious fact that providers of radiocarbon samples carry the burden of probity. Despite the pious hope that misconduct is less frequent among scientists than among lawyers, politicians, and loan sharks, no archaeologist can claim that his or her profession is totally immune from malpractice by those Tankersley (2002) calls "thieves of time."

But this is hardly an excuse for denouncing radiocarbon dating, with its increasingly refined methods of calibration. As long as investigators are free to challenge the findings of their peers, to demand repeatability of results, and to call for tests and test implications, we can expect progress toward understanding the secrets of the past, including the cause of megafaunal extinctions.

EIGHT

KILL SITES, SACRED SITES

It seems we are reluctant to blame our fellow men for
a prehistoric offense against modern conservation ideals
and would rather blame climate or the animals themselves.
The simplest explanation is to attribute all late Holocene
extinction [in New Zealand to] the profound changes
brought about by man with fire, rats and dogs.

**Charles Fleming, "The Extinction of Moas and
Other Animals during the Holocene Period"**

Some archaeologists believe that they can dismiss the overkill theory with
field evidence. They point out that very few New World archaeological
sites have yielded evidence of the killing or butchering of extinct animals,
or even intimate associations of the remains of these animals with early
artifacts. If early humans killed off the megafauna, should there not be
numerous kill or butchering sites, or at least numerous other associations of megafauna remains with contemporaneous human artifacts?
Where is the smoking gun—or, in this case, the smoking spear—to indict
the alleged murderer of the megafauna?

The best evidence for repeated association of Clovis hunters with
extinct large animals involves mammoths. Over 50 years ago Marc Nava-
rette and his father, Fred, residents of the town of Naco on the Arizona-
Sonora border, discovered a remarkable kill site. Weathering out of the
bank of Greenbush Draw on the north side of Naco they found bones
of an adult female mammoth with Clovis points in or near the bones.
Kill sites typically involve butchering but that was not the case here. The
animal appeared to have escaped its assailants. Fresh erosion revealed

the carcass in what archaeologist Ken Tankersley (2002) considers "perhaps one of the most mysterious of the Clovis kill sites." The Navarettes very properly left most of the site undisturbed and called the Arizona State Museum.

Led by Arizona's most experienced archaeologist of the time, Professor Emil Haury, a team from the museum excavated the site. Team members found more well-crafted and perfectly preserved Clovis points made of fine-grained rock. Five were in place and three had been removed by the Navarettes, making a total of eight. There was no doubt that the mammoth had been speared by Clovis hunters. But there was no evidence of butchering or processing of the carcass—no other stone tools or flakes, no hearths, and no charcoal (Haury 1953). Apparently the mammoth, possibly the matriarch of subadults that would be found at the Lehner site and Murray Springs, died out of its attackers' reach. Otherwise they might well have butchered it, or at least recovered their spear points. Clovis hunters reused points when they could, resharpening them if necessary. Unbroken fresh Clovis points are rarely found around a carcass. Unless in the case of the Naco mammoth they remained hidden in an unbutchered part of the body, it is safe to assume the hunters would have retrieved them had they been able. The early flint knappers recognized the best outcrops of cryptocrystalline (hard, fine-grained rock) they needed for working preforms into blades and blades into points. Ideal rock was widely traded; Knife River flint from North Dakota has turned up in Ohio, 900 miles away (Tankersley 2002).

Lacking organic remains, the Naco mammoth could not be dated by radiocarbon. But within a few years three other Clovis kill sites with mammoths and other extinct animals, and one with bison, were found and excavated in dry wash tributaries of the San Pedro River in southeastern Arizona. Immediately to the north of Naco, archaeologists found mammoths associated with Clovis points and other artifacts as well as charcoal. One example is the Lehner site, reported to the Arizona State Museum by Ed Lehner, a longtime resident near Hereford, who saw the Naco site when it was being excavated and realized that he might have something similar on his ranch. There, Professor Haury and his field team found what would eventually prove to be 13 juvenile mammoths with Clovis points, other artifacts, and charcoal (Haury, Sayles, and Wasley 1959). This and other datable sites were estimated to be around 13,000 years old (Taylor, Haynes, and Stuiver 1996), suggesting not only that the undated Naco mammoth was the same age but possibly the matriarch of the animals at the Lehner site. All were *Mammuthus columbi*.

In general, one or a few Clovis points in or near a mammoth carcass, along with other artifacts, are the most archaeologists can expect to find. The Greenbush Draw mammoth is, to my knowledge, unique in being speared but unbutchered. Field evidence of mammoth butchering or processing is also relatively rare, certainly compared to the abundant evidence of hunting and butchering of extinct taxa of bison in the High Plains by people armed with Folsom points (which are a few hundred years younger than Clovis points and have a longer flute). Bones of butchered or processed mammoths are found at kill sites such as Blackwater Draw, New Mexico; Dent, Colorado; Colby, Wyoming; and Murray Springs and other sites along the San Pedro in Arizona. Their absence from Folsom sites suggests that by 12,500 calendar years ago, mammoths were extinct in the western United States. While lacking even one unbroken Clovis point, the oldest Clovis site, Aubrey, in Denton County, Texas, is rich in workshop evidence and is 13,500 calendar years old (Ferring 2001).

Based on what they knew of other Clovis sites in Cochise County, Vance Haynes and Pete Mehringer discovered a particularly informative site, Murray Springs. To the best of my knowledge, Vance and Pete are the only scientists to discover a Clovis butcher/kill site with extinct animal bones. Almost invariably it is amateurs—cowboys, hunters, farmers, hikers—who make the initial discoveries.

Murray Springs is a superb example of a crucial feature lacking in the search for pre-Clovis archaeology: site replication. It displayed many of the major features of the Lehner site in addition to a few of its own.* Between 1966 and 1971, with National Geographic Society grants, Haynes and his field team uncovered and excavated a partly butchered mammoth and discovered fossil proboscidean tracks and bones of eleven young bison in adjoining kill sites. In addition they found stone knives, scrapers, 16 Clovis points (many broken or reduced in size beyond the point of resharpening), and a dire wolf skull, all buried just beneath the Clanton Clay deposit, a "black mat" or stratigraphic marker which blanketed the site like a shroud. Possibly the Clovis hunters guarding their meat cache dispatched the dire wolf, which was attracted by the butchering. The most

---

*Both the Lehner and Murray Springs sites are held by the Bureau of Land Management San Pedro Riparian National Conservation Area, with headquarters in Sierra Vista, Arizona. The bureau can provide information on a self-guided nature trail at Murray Springs (for information consult the nearby San Pedro River interpretive center). A master plan exists for a state-of-the-art interpretive center, but it will only be realized if the public comes to appreciate the importance of Arizona's mammoths and their hunters. The proboscidean ghosts of the San Pedro River have a long way to go to catch up with the interest the public has in the ghosts of the O.K. Corral in Tombstone.

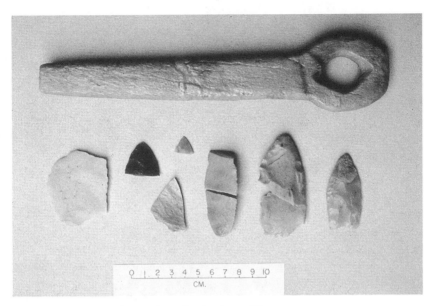

Plate 16. Clovis artifacts. Photo by C. Vance Haynes.

remarkable find was a shaft straightener made of mammoth bone with a hole in one end, like a giant needle. Called *batons de commandant,* these were well known in the Paleolithic of the Old World, but this is the only shaft straightener to have been found with a mammoth at a Clovis site.

Murray Springs yielded a more detailed stratigraphic record than the other Clovis sites. (All are known or thought to be contemporary; see table 7.) Unfortunately, it did not yield a fossil pollen record, which had been of great help in interpreting stratigraphy at the Lehner site. These results intrigued pollen analysts attending the first International Palynological Conference at the University of Arizona in Tucson in 1961. While Pete Mehringer was able to extract pollen at the Lehner site, he could not recover a pollen profile of the environment associated with the Murray Springs mammoth. His pollen profiles at the Lehner site indicated somewhat wetter conditions in the time of the mammoth hunters. Some drying out occurred subsequently, with the fossil pollen revealing the invasion within the last 4,000 years of a Chihuahuan Desert shrub, Indian tea (an *Ephedra* species in the same genus as Mormon tea in the Grand Canyon).

The first Clovis site, rich not only in Clovis points but also Folsom material and younger archaeology, was found near Clovis, New Mexico in the 1930s (Boldurian and Cotter 1999). An apparent frozen meat cache

Plate 16. Shaft straightener made from mammoth bone, found at Clovis site. Photo by C. Vance Haynes.

has been excavated by anthropologist George Frison near Colby, Wyoming (Frison 1998; Frison and Bradley 1999), and a remarkable cache of Clovis blades (much larger than Clovis points and probably their preforms) with mysterious beveled rods of bone embedded in red ochre was discovered in an apple orchard at East Wenatchee, Washington (Mehringer and Foit 1990). The site may be contemporary with an eruption of Glacier Peak in the Cascades; Glacier Peak ash is associated with it. Just possibly, those who left the cache hoped it would propitiate their gods and stop the ashfall. Like hidden treasures, the caches contain materials that were precious to Clovis people, such as red ochre and some of the highest quality tool stone in the Americas, often obtained hundreds

TABLE 7 *Clovis Archaeological Sites*
with Proboscidean Bones and/or Teeth

Site	Taxon and minimum number of individuals	Cultural associations and radiocarbon dates
Blackwater Draw (NM)	mammoth, 6	Clovis lithics 11,630 to 11,640
Burning Tree (OH)	mastodon, 1	11,660; 11,450
Colby (WY)	mammoth, 7	Clovis lithics 11,220; 10,864
Dent (CO)	mammoth, 15	Clovis lithics 11,200 10,670 to 10,980
Domebo (OK)	mammoth, 1	Clovis lithics ~11,000
Dutton (CO)	mammoth, 1	Clovis lithics <11,000
Escapule (AZ)	mammoth, 1	Clovis lithics
Heisler (MI)	mastodon, 1	possible butcher marks 11,770 ± 110
Hiscock (NY)	mastodon, 9	Clovis points 9,150 ± 80 to 11,390 ± 80
Kimmswick (MO)	mastodon, 2	Clovis lithics
Lange-Ferguson (ND)	mammoth, 2	Clovis lithics
Lehner (AZ)	mammoth, 13	Clovis lithics 10,900
Leikum (AZ)	mammoth, 2	Clovis lithics
Lubbock Lake (TX)	mammoth, 2	Clovis lithics 11,100
Miami (TX)	mammoth, 5	Clovis lithics
Murray Springs (AZ)	mammoth, 2	Clovis lithics 10,900
Naco (AZ)	mammoth, 1	8 Clovis points
Navarette (AZ)	mammoth, 1	2 Clovis points
Pleasant Lake (MI)	mastodon, 1	possible butcher marks 10,395 ± 100
Rawlins (WY) (the U.P. mammoth)	mammoth, 1	untyped lithics 11,280
Aucilla River (FL)	mastodon, 1	fluted-point variants, lithics, 33 ivory points

SOURCE: After G. Haynes 2002. Used with the permission of Cambridge University Press.

of miles from the cache site (Tankersley 2002). In fact, apart from mammoths, it is remarkable how rarely the fossils of extinct Quaternary animals are found in thoroughly convincing association with prehistoric human artifacts (see table 7). The lack of associations is especially noteworthy because Clovis points are well represented in the fossil record, particularly in the plow zone of the Midwest.

To date no Clovis points or other early human artifacts are known to be associated with ground sloth remains. One famous claim of such an association at Gypsum Cave (Harrington 1933) was based on stratigraphy and has not been verified by radiocarbon dates (Heizer and Berger 1970). The wooden darts or shafts from Gypsum Cave yielded radiocarbon dates at least 8,000 years younger than the sloth dung in which they were embedded. No prehistoric artifacts have, to my knowledge, been found in Rampart Cave. (Clovis artifacts are rare in caves in general, and none have been discovered in any Grand Canyon caves.) The fossil record thus yields no direct evidence of hunters having killed ground sloths, much less of young people using the animals for target practice, as I have suggested. Convincing camel kill sites also have yet to be discovered. (As discussed in chapter 7, hopes for such a site at Tule Springs were dashed.) The many southwestern fossils of horses and mastodons found to date also have no Clovis or other archaeological associations. Only rarely does extinct North American megafauna turn up in archaeological sites. Before seizing upon that fact as evidence that humans did not kill off American proboscideans, it is worth noting that "*The United States contains more megamammal killsites than there are elephant killsites in all of Africa—a land mass that is much larger than the United States*" (G. Haynes 2002b, 183; emphasis in original).

The record regarding bison is somewhat different. Clovis hunters clearly hunted bison at Murray Springs and other sites in the West, leaving their points with the bones. Overall, there are far more associations of human hunters with bison than with mammoths or mastodons. The vast majority of bison kill sites, however, are younger than Clovis and feature Folsom points, especially in the High Plains to the east of Arizona and the Rocky Mountains. Some archaeologists take this as a further argument against overkill. As my archaeologist friend Jim Hester put it in about 1965, voicing a view widely held by archaeologists, "How can you invoke overkill to explain extinction of the mammoth when . . . we have many more younger sites . . . associated with bison kills? They stretch over thousands of years, from Folsom on to Midland. In contrast, there are only a few Clovis associations with mammoth. Yet the bison survived and mammoths are gone. If people did it, the field evidence is backwards." The answer to Jim's question is important. In the fossil record a catastrophic event that results in extinction may not last long enough to accumulate appreciable evidence. If hunting is sustained over thousands of years, a great deal of field evidence of past bison hunts can be expected, but if other plant and animal resources are limited, bison as the only large and widespread prey is not as vulnerable to extinction. In this "arms race," bison become wilier at avoiding hunters, who in turn learn better hunting techniques. After European contact in the late 1400s, disease swept through Native American populations. In a century bison began to spread into the southeastern United States and into New York State, where they had not been for thousands of years. Their numbers dropped again as European settlers began to hunt them and the Anglo frontier pushed buffalo back west.

As for cave paintings, according to Gary Haynes, "There are no known cave paintings, portable artwork, carved figurines, or petroglyphs that *clearly and unambiguously* portray Clovis-era images" (G. Haynes 2002b, 158). In contrast, extinct mammals, especially mammoths and woolly rhinoceroses, along with living species, are magnificently portrayed in the artistic galleries drawn by Paleolithic people in Europe. They are depicted not only on rock shelter or cave walls and ceilings but also on ivory or bone artifacts. Such ancient art displays superb knowledge of animal behavior and morphology. For example, a female reindeer is portrayed sniffing a newborn, as reindeer and caribou do to determine their infants' parentage (Guthrie 2005). Some of the finest drawings, such as those in Chauvet Cave in France (Chauvet, Deschamps, and Hillaire 1996), are over 30,000 years old. With the exception of a widely reproduced rock drawing of a putative proboscidean in Utah and perhaps one other, there is nothing in the New World to suggest a lengthy association with mammoths and other extinct species. American rock art portrays living animals only and is considered mainly postglacial, and therefore post-Clovis, in age.

Cultural associations with extinct animals have also been difficult to find in other areas outside Afro-Eurasia, the continents of human origin. Australia's Willandra Lakes region, for example, is rich in archaeological remains, especially those of Holocene age, and in paleontological sites with extinct fauna. To my knowledge, however, there is no overlap between the two, though the human presence at Lake Mungo goes back over 30,000 years. At Lancefield, in the state of Victoria, an incredibly rich deposit of 10,000 extinct kangaroos is associated with charcoal originally radiocarbon dated at 26,000 years and now considered older. Investigators Richard Wright and Dizzy (Richard) Gillespie found no decisive cultural association and no indisputable evidence of butchering of the kangaroos. Similarly, few definite cultural associations with extinct species are found in Madagascar. The same is true of the Mediterranean islands and Pacific islands, including Hawaii, where the Polynesians arrived only about 1,500 years ago (Martin and Steadman 1999, 24, 26).

David Steadman, Greg Pregill, and Dave Burley (2002) have described a unique association at the Tongoleleka site on the island of Lifuka in the Tongan archipelago. There, bones of an extinct megapode or bush turkey, *Megapodius alamentum,* occur adjacent to those of a giant extinct iguanid *(Brachylophus,* undescribed sp.) as well as pottery of the pioneering Lapita culture and bones of introduced chickens *(Gallus).* The associated fossils are virtually the same age: weighted mean average of calibrated radiocarbon ages in calendar years are 2,780 to 2,750 for six dates on the chicken bones, 2,840 to 2,760 for six dates on the iguana, and 2,950 to 2,780 for eight dates on the megapodes. This is a fabulous find, nailing the association of extinct animals with human activity, exactly what is not easy to uncover 10,000 years earlier in Clovis time in North America. Perhaps, with time, more such associations will be found, although very few have turned up in a quarter century of remarkable success at discovering new taxa of insular fossils, many of extinct land birds.

Only on the South Island of New Zealand are prehistoric human artifacts abundantly associated with extinct fauna. There the overlap between evidence of human occupation and bones of 10 species of moa teetering on the edge of extinction covers less than 100 years (Holdaway and Jacomb 2000; Worthy and Holdaway 2002).

As far as direct archaeological evidence is concerned, therefore, the most we can say for the Americas is that Clovis hunters overlapped with mammoths and bison and at least on occasion they hunted both. If human artifacts are not found with bones of ground sloths, horses, and extinct camelids, and rarely with mammoths, why should archaeologists pay attention to the view that people forced the extinctions (Meltzer 1993)? Zooarchaeologist Don Grayson, for example, concludes from the lack of camel kill sites that people rarely or never killed camels (Grayson 1991). Even if the extinctions did happen in Clovis time, perhaps Clovis hunters had little or nothing to do with them. After all, fewer than 70 mammoths are all that can be accounted for at some 20 known Clovis sites (see table 7). The number seems trivial in terms of the loss of a continental population not only of mammoths but also of many other large mammals that must have totaled many millions. Nevertheless, for three reasons I would suggest that the lack of kill sites is actually supportive of the overkill model.

First, if there are few contemporaneous associations of extinct megafauna with Clovis artifacts, there are fewer with post-Clovis artifacts. Although Folsom points, for example, are only a few hundred years younger than Clovis points, archaeologists have not found them in a clear-cut association with any extinct species beyond taxa of bison. There are no unambiguous records of mammoths, horses, camels, ground sloths, or other extinct genera of mammals anywhere in North America younger than those of Clovis time. A camel (terminal date 10,080 ± 179) once thought to be associated with Folsom artifacts is actually not well associated. In fact, the more usual pattern is an older layer of extinct megafaunal remains with few, if any, artifacts, plus a younger layer with much evidence of people and what they ate, but no extinct species (Martin 1986).

At a minimum, then, the archaeological evidence indicates that by the time Folsom points came into use, about 10,700 radiocarbon years ago, most or all of the great mammals were already gone. Except for extinct taxa of bison and possibly California Condors in West Texas, there were no Holocene (postglacial) mammoths, ground sloths, or other extinct megafauna. Similarly, Australia (45,000 to 55,000 years ago) and Madagascar (500 BC to 1500 AD) have revealed few, if any, cultural artifacts of pioneering peoples in association with remains of megafauna on the edge of extinction (Martin and Steadman 1999, 40).

Second, even if spears were used to kill, for example, ground sloths, stone spear points would not necessarily be found in association with Shasta ground sloth bones. They would have been recovered and used again. This may be a partial explanation for the paucity of apparent kill sites of animals other than bison. As for the number of bison kill sites, perhaps these are attributable in part to the fact that bison were commonly killed in a herd or group. Or perhaps by Folsom time the location of quality stone for making tools was so well known that much less care was taken to recover points than in the days of the mammoth hunters.

Third, the rate of extinction would have determined the archaeological visibility of the event. An event that occurs in a very short period (a few tens or hundreds of years throughout a continent, as I posit for the American extinctions) will scarcely be detectable in the fossil record (Martin 1973; Mosimann and Martin 1975). Rapid and massive extinction of large animals, for whatever reason, would have left virtually no trace. I believe the impact of the first human invaders was so sudden and severe, and the opening and closing of the Clovis window so rapid, that we may never find many kill sites. Indeed, I believe we are fortunate to have any (G. Haynes 2002b), especially in open areas where sedimentation (and hence a fossil record) is scant. If a stratigraphic alignment of the first human arrivals with the last presence of an about-to-be-extinct fauna is difficult to detect only 3,000 years ago on a small island like Lifuka in the Tongan archipelago, it is small wonder that few mammoth kill sites 13,000 years old have turned up in America and there are as yet no megafaunal kill sites in Australia (some consider Cuddie Springs to be an exception; Wroe and others 2004). Only under unusually favorable circumstances does a very careful excavation reveal the field evidence archaeologists have long demanded.

This also explains the lack of art: the animals vanished while the Clovis people were on the move. The lack of any artistic depictions of these animals is an argument in favor of a very short temporal overlap between them and the first hunters. In the Old World, where the overlap was much longer and the extinctions fewer and more gradual, art is abundant.

In addition, quite apart from the overkill model, it is always particularly fortuitous to discover creatures that died and fossilized right at the moment of extinction, especially if that extinction was almost instantaneous, and especially for taxa that were relatively rare to begin with. The most we can expect is some chronological control on when the last populations were alive. For mammoths, horses, ground sloths, and others in western North America, we may have the right millennium of extinction (between 13,500 and 12,500 calendar years or 11,500 and 10,500 radiocarbon years). By leaning on negative evidence, we may narrow the chronology down in at least a few cases, such as those of the Columbian mammoth, the Shasta ground sloth, and Harrington's extinct goat, to what may prove to be the right century or two (around 10,900 to 11,100 radiocarbon years). If all 32 extinct North American genera were struck down within a few decades, the event is beyond resolution by current dating techniques.

Those who deny overkill often turn to natural causes to account for the extinctions. Such an approach has major problems. Detailed fossil pollen and macrofossil plant records are continually appearing for habitable parts of the planet and various dating methods, especially radiocarbon dating, are available for chronological comparisons. If large mammals thoughout North and South America disappeared simultaneously around 11,000 radiocarbon years ago as a result of some extraordinary climatic shock, nothing similar is apparent outside the Americas, not even in the West Indies. Field evidence of climatic or other forcing of megafaunal death is even harder to uncover than kill sites. Where are the "freeze sites," if some cold shock wave is involved? Would such a shock kill off plateau- or montane-ranging species before they could descend to adjacent lower and warmer elevations, like the Mojave Desert in Arizona and California, watered by the Colorado River? In any case, to match the global "deadly syncopation" of Ross MacPhee, the killer climatic change would have to have struck Australia long before the Americas and Madagascar, New Zealand, and the Pacific islands long after. If "killer cold" exterminated large mammals in South America around 10,500 radiocarbon years ago, might we not expect some concurrent losses of megafauna in Australia, South Africa, and New Zealand?

Most fossils are proof of nothing more than an organism's existence and death. In very few cases can we infer an unambiguous cause of individual mortality, much less of its extinction as a species. This makes it difficult to extract definitive answers from the fossil record. Those who favor the climate theory can argue that the scarcity of kill sites indicates that people had little to do with the extinctions. Those who favor overkill can argue that a mere decade of human impact in any new region would have been enough to entrain the extinctions but highly unlikely to be reflected in the fossil record (Mosimann and Martin 1975; Mithen 1997).

The case of the dinosaurs makes an interesting comparison here. If the main evidence for dinosaur extinction were fossils alone, few if any paleontologists would have dreamt that that extinction was sudden. Even with the discovery of the Chicxulub Crater in the Yucatan, which coincides in age with the Cretaceous-Tertiary boundary and makes the case for a catastrophic extinction of Mesozoic biota, some paleontologists have been slow to abandon their gradualist views, and not all have done so. The Alvarez model implies that the dinosaurs all died out in less than a year at the end of the Cretaceous, as the result of major disruption of the atmosphere following a severe extraterrestrial accident, such as an asteroid impact. Many geologists regard such a conclusion as robust. But sections of continental deposits dating from the end of the Cretaceous that could potentially harbor the last bones of non-avian dinosaurs are few. The most common dinosaur in the fauna, *Triceratops,* is, as one would expect, the one found closest to the boundary. Dinosaurs simply are not common enough to trace easily to the boundary. The same may be anticipated for at least some of the 32 genera of mammalian extinctions in the Americas in near time. As paleontologists have discovered, the fossil record may not be good enough to incorporate all of the species lost when a catastrophe occurs.

In the Southwest, twig figurines are particularly likely to be associated with extinct animal remains. The associations are not contemporaneous, so they tell us little about the extinctions, but they do shed interesting light on past human attitudes toward the extinct beasts. I first began focusing on these associations in 1969 at Stanton's Cave. Bob Euler's particular interest in the cave centered on the numerous split-twig figurines that he and others found there. Each figurine was 4 to 6 inches in length and had a head, neck, and legs attached to a body, all constructed from a single willow twig. The figurines looked like some kind of small ungulate, either a mountain sheep or a deer (see plate 17). Some even had a slender twig or a splinter inserted through the midsection, very likely symbolizing a spear.

Bob's National Geographic Society grants to investigate the cave were triggered in part by rapid loss of these figurines from it. By the time his excavations began, at least 75 figurines had found their way to museums, some via well-intentioned individuals who sought to keep them safe. The problem was that both the "rescuing" and the looting destroyed context. With no knowledge of the figurines' provenience (exact location within the cave), a crucial piece of information in site analysis, it was impossible to reconstruct any pattern that might help determine their meaning or function. Undoubtedly any figurine arrangements originally left on the cave floor had been the first to disappear. However, Bob's team discovered some figurines in clusters of up to five that had been carefully cached between flat rocks by human hands. There were a total of over 160 figurines, most of which radiocarbon dated at 3,000 to 4,000 years.

Interestingly, Stanton's Cave showed no evidence of ever having been lived in, despite its apparent suitability. That seemed strange. Bob and his field team discovered no hearths, no kitchen middens containing bone scraps, no stone knives or scrapers—not even any potsherds. A few domestic artifacts did appear, but only at or just beneath the floor of the cave. These scraps were probably associated with the Pueblo II (1050 to 1150 AD) Kayenta Anasazi ruin on a terrace at the mouth of South Canyon, a few hundred yards upstream. They were not associated with the figurines.

Although I do not pay much attention to prehistoric artifacts, the

Plate 17. Split-twig figurine from Stanton's Cave. Photo courtesy Arizona State Museum.

figurines haunted me. Surely they had symbolic value. What were they made for? Why were so many found in Stanton's Cave? Had they some-how been associated with the abundant remains of Harrington's extinct goat, which were found in quantity in the cave and were at least 6,000 to 30,000 years older? If there was a connection—if, for example, the figurines had intentionally been placed near the remains—did their makers believe that these remains were of an existing creature, or did they recognize that they were of something different, something extinct?

Bob suggested the former, and raised the possibility that the cave had been a special place, a sacred site: "It is generally agreed that [the figurines] represent some form of magico-religious ritual, the twigs inserted through the body having been representations of spears that functioned to insure success in the actual hunt" (Euler 1984, 9). I wondered if the figurine makers hoped to find some of the mysterious animals, the extinct goats, still alive. Perhaps they imagined that they might resurrect them. For his part, Steve Emslie thought these people were expert hunters and trackers who immediately knew the remains were not of living animals but of ancestral ones (personal communication, November 2001). Their presence would have made a cave—a symbolic or accepted entrance to the total darkness of their underworld and the ancestors—a sacred place, a place for shrines.

The mystery was clarified when Emslie discovered an unvandalized cave in the Grand Canyon rich with figurines associated with the much older remains of extinct animals. He called the new site Shrine Cave (Emslie, Euler, and Mead 1987). His team also discovered five other relatively undisturbed caves high in vertical cliffs of the Grand Canyon. Emslie reported, "Radiocarbon dates indicate cultural use of these sites between 4300 to 3700 B.P. . . . Unlike most other Archaic sites in the Grand Canyon, these contain numerous rock cairns as well as cairns built partially or entirely of indurated packrat-midden fragments of late Quaternary age. Archaic artifacts, including split-twig figurines, appear to be deliberately associated with fossil material of extinct mountain goats and other vertebrate remains" (Emslie, Mead, and Coats 1995). In Gypsum Cave, Harrington (1933) reported an intimate association of Archaic artifacts with extinct fauna. Although he thought they were contemporary, as we have seen, radiocarbon dates indicated otherwise. Judging by the quality of the artifacts, including painted throwing sticks, Gypsum Cave could have been the most sacred sanctuary of all, a place where Archaic people venerated animals extinct for thousands of years.

In the Grand Canyon, Archaic habitations or campsites are scarce. Perhaps the rugged landscape that made the canyon unsuitable for dwelling enhanced its value for ceremonial purposes. To date, hundreds of split-twig figurines and dozens of cairns have been found in Grand Canyon caves, many of them difficult of access. The sheer number of these apparently sacred sites is remarkable given the paucity of dwelling places in the canyon at the time the figurines were created, or indeed at any time prior to widespread Pueblo II–age occupations.

Twig figurines of animals have also been recovered from Archaic sites in Nevada, Utah, and Colorado. These are mostly younger and probably served a different function, as they are associated with habitation sites, rather than with extinct animal remains. Those from Cowboy Cave in Utah, for example, while the same age as those from the Grand Canyon and westward, are less carefully made and appear more secular than sacred. They are associated with a variety of other artifacts, including Gypsum Cave points, presumably used in hunting deer and mountain sheep.

The split-twig figurines could tell us nothing concrete about the extinction of Harrington's mountain goat in the Grand Canyon. They were thousands of years too young to have had anything to do with the extinct animals of the Pleistocene. This did not mean, however, that they were completely tangential to the question. The positioning of cairns, twigs, or effigies near dung or bones of the extinct goats was ample evidence of prehistoric people's inordinate interest in these goats (an interest shared by investigators such as Larry Coats, Steve Emslie, Bob Euler, Dick Harington, Jim Mead, and Norrie Robbins). For whatever reason—a focus on the hunt, or simple fascination—our ancestors were strongly drawn not only to living large animals, but also to the remains of extinct ones. Occasionally, extinct animal remains are found in much younger archaeological sites, as at Casas Grandes in northern Chihuahua. There the "rock shop," a small room about 1,000 years old, contained many teeth of mammoths, Paleozoic fossils, and semiprecious stones (DiPeso, Rinaldo, and Fenner 1974).

The intense interest of prehistoric people in Quaternary fossils is especially evident in the Mediterranean region. Adrienne Mayor (2000) credits Austrian paleontologist Otto Abel with the idea that the Cyclops of Homeric times may have originated in the discovery in caves in Sicily of crania of fossil dwarf mammoths: their medial nasal cavity suggests one large eye socket. On a similar front, a major preoccupation (approaching an obsession) with depicting large mammals appears 30,000 years ago as an important component of Old World Paleolithic art (Guthrie 2005).

When I started reviewing literature for this chapter, especially Emslie, Mead, and Coats (1995), I was surprised by the strong suggestion of "megafauna worship" by Archaic people. There may be a sociobiological reason for this attraction. For at least two million years in Africa, the interaction between evolving humans and large animals would have been dynamic, intense, and durable. A deep emotional reaction to large animals would have been incorporated into our genome. Our Pleistocene ancestors hunted or scavenged large mammals, and some of them—lions, hyenas, leopards—returned the favor. It would not surprise me if some residue of these ancient terrors and traumas lingered in our dreams.

NINE

MODELS IN COLLISION

Climatic Change versus Overkill

I will lay aside impartiality. I think the overkill theorists have
the more convincing argument for what happened in America
10,000 years ago. It seems that the Clovis people spread
through the New World and demolished most of the large
mammals during a hunters' "blitzkrieg" spanning several
centuries.

E. O. Wilson, 1992

Despite accumulating evidence that humans caused the megafaunal extinctions, some members of the climate-change school are in deep denial
(Grayson and Meltzer 2002, 2003). A cadre of archaeologists, especially
those who claim or prefer to believe that people were in the New World
before the extinctions began, agrees with them. In addition, many vertebrate paleontologists of my generation, born in the first third of the
twentieth century, support the climatic paradigm.* Although details of
the process are rarely available, climatic change is often the answer to
the question of what accounts for all the numerous extinctions lacing the
fossil record.

---

*Vertebrate paleontologists who have supported climatic extinction models, at least in
the past, include Elaine Andersen, Tony Barnosky, Russ Graham, Don Grayson, John Guilday, Dale Guthrie, Claude Hibbard, Ev Lindsay, Ernie Lundelius, Larry Martin, Jeff Saunders, Bob Slaughter, and Dave Webb. For the Martin-Grayson debate, see "Clovesia the
Beautiful," pp. 39–64 in Russell 1996.

The overkill model remains controversial, to say the least. (My department head once inquired, in what I hoped was a friendly tone, "Hey, Paul! How far out on that extinction limb do you think you can go?") On the other hand, while those archaeologists who pin their faith on the existence of pre-Clovis human populations in the New World are especially prone to oppose the concept (for example, Grayson and Meltzer 2003), one or another version of an overkill model finds traction among researchers active in many fields.*

Proponents of various views, including climatic change and overkill, are chapter authors in Martin and Wright 1967 and Martin and Klein 1984; the debate has been with us for some time. It points to some fundamental differences in methodology, outlook, attitude, and training that could be as illuminating as the final answer to the extinction controversy itself. In some cases the approaches taken may be personal, based on views or agendas that go decidedly beyond the narrow issue of extinction and its causes. And some researchers favor a mixture of climate change and human hunting as the functions forcing extinctions in near time. Their approach is politically less risky, but also less testable. I believe it is redundant. If the climates are always changing, climatic change is inescapable. And as John Alroy (1999) and Lyons, Smith, and Brown (2004) show, size made all the difference in terms of which animals went extinct.

Many who support the climate-change theory, like my one-time committee member Claude Hibbard, are vertebrate paleontologists who have thousands of extinctions to account for in the Cenozoic, the majority long predating the time of *Homo sapiens*. In northern Eurasia and especially in North America, both range changes and animal extinctions occur during or close to the time of the Younger Dryas cold snap, recognized widely at least in the Northern Hemisphere. In addition, archae-

---

*In anthropology these include Fiedel 2003; Fiedel and Haynes 2003; G. Haynes 2002a, 2002b; Mithen 1997; Porcasi, Jones, and Raab 2000; Redman 1999; Waguespack and Surovell 2003. In addition, a partial scan of other disciplines reveals the following publications that elaborate on or apply the model or theory of overkill. The response involves the following fields: biogeography (Brown and Lomolino 1998; Pielou 1991), conservation biology (Kay 1994, 1998; Kay and Simmons 2002), cultural studies (Sayre 2001), economics (V. Smith 1975, 1999), history (Flores 2001; Sheridan 1995), geochemistry (Gillespie 2002), historical ecology (Burney 1993; Diamond 1992, 1997; Lyons, Smith, and Brown 2004), holistic management (Bonnicksen and others 1999; Burkhardt 1996), marine biology (Dayton 1974; Simenstad, Estes, and Kenyon 1978), paleobiology-paleontology (Alroy 1999, 2001; Azzaroli 1992; Boulter 2002; Fisher 1996; Flannery 2001; Holdaway and Jacomb 2000; Merrilees 1968; MacPhee and Marx 1997; Raup 1991; Steadman 1995, n.d.; Ward 1997), popular works (Barlow 2000; Lange 2002; Leakey and Lewin 1995; Wilson 1992), and zoology (Coe 1981, 1982; Janzen 1983; Owen-Smith 1989, 1999).

ologists note that there are no more than a few kill sites of large extinct mammals in North and South America.

Certainly climate changes can reduce, increase, or shift species' ranges; reduce or increase the availability or nutritional quality of forage; change the length of seasons; and otherwise regulate animal populations. Overall, however, the climate-change proponents seem to me to assume their conclusion rather than to prove it. Just saying that "climates change and cause extinctions" does not make it so. In any case, climatic change is invoked whenever it is necessary to explain an otherwise mysterious extinction event, and paleontologists have lots of those. In addition, talk of climate change has the advantage of not distressing those concerned with cultural sensitivity—some racial groups could suffer bad press if the word gets out that their ancestors might have helped to exterminate moas, *Megalania*, mammoths, or megathcriums. This may be a much more serious problem for social scientists than Earth scientists, although all aspire to cultural sensitivity. In this chapter I will outline some of the climatic evidence, showing why it offers little if any help in accounting for the extinctions in near time. Historically, droughts like the ones in Kenya and other parts of Africa have had a huge impact on numbers of large herbivores. So have poachers. So has disease. The climate is always changing, so we can always bring it into the theater of explanations. I claim that human invasion of empty lands, a unique event in different parts of the world in near time, is an overriding variable that explains much more than these other factors.

Any plausible climatic explanation for these extinctions must meet three criteria. First, the evidence must show that there in fact was significant climate change around the various times of the extinctions in the various places where they occurred. Second, the change (or changes) must, alone or in combination with other factors, have been unique in the Quaternary. A change closely resembling others that the megafauna had repeatedly survived, like drought, is not a good candidate in the search for explanations of extinction. Third, the change must have been one capable of striking large terrestrial mammals while sparing most other terrestrial animals, as well as plants and marine life. Unfortunately for its proponents, the climate-change theory meets none of these criteria.

To take the most basic first, the record simply does not show significant climate change spreading across the globe in synch with the time-transgressive extinctions of near time, which, as we have seen, progressed from Australia to the Americas to oceanic islands over a period of roughly 50,000 years.

One change frequently posited to have caused the extinctions is known as the Allerød–Younger Dryas oscillation. As revealed in ice cores, fossil pollen diagrams, and fossil beetle faunas, around 15,000 to 12,000 years ago, the climate of western Europe suddenly switched back to glacial cold (known as the Younger Dryas) after a few thousand years of almost interglacial warmth (the Allerød). For a millennium at most, in Scandinavia during the Younger Dryas glacial ice and periglacial tundra readvanced. (A member of the rose family, *Dryas* is one of the tundra herbs preserved in late-glacial silts and clays.) Then there was a rapid reverse switch to postglacial warmth, which saw the spread of woodland over the formerly treeless periglacial landscape in Europe, Siberia, and the eastern United States. In brief, the evidence for important vegetation change, as I know it, is strongest in western Europe, where those digging for peat as fuel revealed stratigraphic sections of warmth-loving plants below and above a clay unit with *Dryas octopetala* and other tundra herbs that supported the argument for a reversal to glacial cold. Nothing this dramatic can be detected in the western United States, where thousands of packrat middens indicate climatic change from cold to warm, without an obvious reversal.

Even in the Old World, where its climatic signal is strongest, the Younger Dryas is not closely tied to sweeping megafaunal extinction. In the arctic and subarctic, the last of numerous records of woolly mammoths, for example, postdate the Younger Dryas by thousands of years and are not aggregated in a way suggesting simultaneous extinctions (MacPhee and others 2002). In their search for ancient pathogens in frozen tissues or well-preserved bones in the Siberian permafrost of the Taimyr Peninsula, various investigators, both Russian and foreign, have accumulated a wealth of new radiocarbon dates on Eurasian megafaunal extinctions and range changes.

For that matter, it is not clear that the Younger Dryas could have forced *any* European or Russian megafaunal extinctions. For some time, zoologists viewed the Younger Dryas as having forced the extinction of the Irish elk or giant deer, *Megaloceros,* which vanished from Ireland, long thought to be its last refuge, around the Younger Dryas (Barnowski 1986; Stuart 1991). Recent fossil finds, however, complicate the story. Giant deer have now been dated to 9,200 radiocarbon years on the Isle of Man and to 9,400 radiocarbon years in Scotland, 1,400 years later than the climate-driven cold millennium of the Younger Dryas (Gonzalez, Kitchener, and Lister 2000). In the subarctic, woolly mammoths and musk oxen lasted for thousands of years after the Younger Dryas (MacPhee and others 2002).

What about the New World? Here, the paleobotanical evidence for a late-glacial readvance of tundra 11,000 radiocarbon years ago is less apparent than in the Old World. The Allerød–Younger Dryas shift simply may not have been as sharp here, particularly in the arid Southwest. Some regions, such as Arizona and adjacent states, show gradual, not sudden, transformation from glacial to postglacial types of upland vegetation (Martin 1999). In western North America both fossil pollen (Anderson and others 2000; Weng and Jackson 1999) and fossil packrat midden records (Betancourt, Van Devender, and Martin 1990) fail to indicate a sudden sharp switch from warm to cold conditions comparable to the Allerød–Younger Dryas switch in western Europe. In South America, neither microhistology nor pollen analysis of ground sloth dung indicates significant change in local plant ranges around the time of the extinction of the Shasta ground sloth. Though the climate changed more dramatically in northern North America, the number of extinctions was no higher there (Monastersky 1999).

In addition, while the Allerød–Younger Dryas switch at least coincided chronologically with the megafaunal extinctions in continental America, it was out of synch with those on other landmasses. It occurred much too late to account for the extinctions in Australia, Europe, and much of Asia, and left no imprint on the oceanic islands. Unlike the time-transgressive prehistoric arrival of human colonists, climate changes do not, on a global scale, provide a close fit to severe near-time episodes of extinction. The Australian large-animal extinctions, for example, are 30,000 years older than those in the Americas. It is difficult to see how any single climate change could have caused them both, and indeed there is no evidence of such a change; Australia had neither glaciers nor rapid postglacial warming (Ward 1997, 152). On both landmasses, however, the extinctions coincide closely with the widely disparate times of human arrival.

The West Indies provides another test. As discussed above, the fossil record suggests that at least 4,000 years after megafaunal extinction in North America, Cuba and Haiti still harbored diminutive endemic ground sloths. Evidently, then, the hypothetical climatic catastrophe wiping out the full-size ground sloths in both North and South America will not account for the much later extinction of the small Cuban-Haitian sloths (though it is hard to imagine how they could have escaped). A consistent theory would have to invoke a second climatic or environmental crisis that selectively swept away dwarf Cuban and Hispaniolan ground sloths thousands of years later, at a time when few if any extinctions of animals of any size are known on the mainland.

Finally, no global model of large-mammal extinctions corresponding to the Younger Dryas or other cold reversals or cold stages recognized in the Northern Hemisphere yields examples in New Zealand. The last glaciation, known in New Zealand as the Otiran, affected climates and vegetation on both North Island and South Island, and the postglacial warming did so as well. The habitats and ranges of moas shifted accordingly. Nevertheless, there is no record of extinction of moa species until about 500 years ago (Bunce and others 2003). (An earlier extinction episode struck the terrestrial fauna of New Zealand around 2,000 years ago, when Pacific rats arrived; see chapter 6.) If a climatic catastrophe forced extinctions in Australia around 46,000 years ago, and in North and South America 10,000 to 12,000 radiocarbon years ago, neither event registered in New Zealand. The avian extinctions accompanying colonization on eastern Pacific islands also occurred thousands of years after the Younger Dryas and other major late-glacial climate changes (Martin and Steadman 1999). They appear to coincide with Polynesian landfall, sometimes in two stages.

The climate-change model fails not only the test of time-transgressiveness in the near-time extinction pattern but also that of uniqueness. Quite apart from the Allerød–Younger Dryas switch, the climate was indeed changing in various parts of the world throughout the Quaternary. There is nothing unusual in that. Proxy climatic data from ice cores and other sources show that ice age climates were continually and rapidly fluctuating; the late Quaternary was a flickering switch of climatic change. Indeed, near time and the Quaternary have been so variable that the expression "climatic change" is redundant (Hughen and others 1998).

Climatic changes earlier in the Quaternary apparently equaled those of the late Quaternary in amplitude (Porter 1989). For example, Greenland ice cores show that climatic changes like the Allerød–Younger Dryas–Preboreal switch, from warm to cold and back again, had repeatedly occurred earlier (Dansgaard and others 1993; Alley 2000). The displacements downward and southward of many species during the last glacial age reflect perhaps a drop of six to nine degrees Fahrenheit (3.3 to 5 degrees Celsius) in mean annual temperature and a decrease in carbon dioxide. These are evident in the fossil record from ice cores, marine cores, pollen diagrams, and changes in sea level due to the expansion or contraction of glacier volume. By 8,000 to 10,000 radiocarbon years ago, most of the present climatic gradient was back in place. The ice cores (Alley 2000) show that over 100,000 years there were more than two dozen reversals like that of the Younger Dryas, with very rapid changes of at least 18 degrees Fahrenheit (10 degrees Celsius) from warm to cold and cold to warm.

Those earlier changes, however, were unaccompanied by extinctions of large mammals, which we may assume evolved to deal with a wide range of climatic variation. New World equids, camelids, and proboscideans, for example, proliferated over tens of millions of years of constantly changing climates. Even the small ground sloths and other insular species of the West Indies would have been selected for survival under widely varying climatic conditions, otherwise they too would have gone extinct during the start of the ice age, long before near time.

The range of some large animals lends further support to this conclusion. Modern-day elephants, for instance, occupy many climate belts in Africa, from the Skeleton Coast and dunes of Namibia to the rain forests of the Congo and the savannas beneath Kilimanjaro. The woolly mammoths occupied northern Eurasia and northern North America; the Columbian mammoth's range was transcontinental, from Alaska south throughout most of the United States, and went from an elevation of 9,000 feet in the mountains of Utah to sea level in Florida and Mexico. It seems unlikely that such adaptable animals could have been totally wiped out by even the most severe weather conditions. Indeed, there is some direct evidence that the climate was not a problem for them; in one case the tusk growth of a mastodon from Michigan suggests favorable environmental conditions close to or at the time of mastodon extinction in that region (Fisher 1996).

Given the large mammals' demonstrated ability to survive the "normal" severe climate swings of the Pleistocene, those favoring climatic change as the primary cause of the extinctions must invoke some unique event (or time-transgressive series of events). Unfortunately for them, nothing unique jumps out from the wildly fluctuating changes reflected in the marine and ice cores of Greenland and Antarctica in near time (Dansgaard and others 1993; Alley 2000). A global catastrophe, something off the scale, would surely have registered in these ice cores. In addition, we should see such a catastrophe in the fossil record, and we do not. The last shift from a glacial to an interglacial climate saw the last major change in the regional distribution of desert plants and animals. However, rather than evidencing a sudden, widespread shift in vegetation types, which might reflect some sort of unique climatic crisis, the midden record changes gradually, over thousands of years, as species favoring cooler climates are replaced by those favoring warmer ones (Betancourt, Van Devender, and Martin 1990; Martin 1999). In mountainous country the late-glacial-age shifts of 2,000 to 3,000 feet shown by fossil pollen and midden records could have been traversed easily by large herbivores in less than a week, if not a day.

In brief, while the environment south of the ice margin fluctuated and at times changed dramatically, there is nothing to suggest that those changes alone could have forced, for example, the extinctions of over 30 genera of large mammals in North America. The best negative evidence comes from the repeated sudden, severe climatic switches, unaccompanied by extinctions, seen in over half a million years of change in ice cores. There were many such switches beginning long before extinctions struck.

The taxonomic selectivity of the late-Quaternary extinctions poses a third problem for the climate-change model. With one possible exception, no tree or shrub extinctions are evident in the near-time fossil record. In the western United States, rich in well-dated fossil deposits of extinct large mammals, no contemporary plant extinctions are reported. The recent discovery and description of an extinct and a highly displaced species of spruce, *Picea critchfieldii*, from the southeastern United States (Jackson and Weng 1999) is a reminder of how unusual plant extinctions have been in North America since the mid-Cenozoic. In contrast, extinctions or extirpations of numerous plants, including temperate genera of trees, occurred in the Pliocene and early Pleistocene of Western Europe. These are thought to reflect climatic change at the end of the Tertiary, when temperate species were barricaded by the Alps from any easy retreat southward during times of glaciation and the spread of tundra-taiga south to the Alps (Leopold 1967).

Birds, small terrestrial mammals, marine mammals, and most beetles were much less vulnerable to whatever caused the extinctions of the large terrestrial mammals. Beyond scavengers of large mammals, the extinction wave left no imprint on the fossil record of birds or small mammals, both well represented in Stanton's Cave and many other deposits in the Grand Canyon and the Colorado Plateau. There were no marine extinctions in near time, though these typified mass extinctions in deep time. On islands off California, pinnipeds did suffer local depletions when their rookeries were hunted out (Porcasi, Jones, and Raab 2000). This depletion was episodic, not systemic, and the seals and sea lions recolonized after human hunters left to seek resources elsewhere. The rich fossil record of Quaternary beetles is sensitive to climatic change but lacks much evidence of extinction. The few extinct genera (Coope 1995) are mainly of coprophiles (scarab or dung beetles), suggesting coextinction of scarab beetles and megafauna. That is, when megafauna declined globally, especially in the Americas and Australia, there was a major reduction in dung deposition and thus a major shrinkage of dung beetle habitat.

Large animals are more vulnerable to hunting than smaller ones—they are generally easier to track, locate, and spear, and they reproduce more slowly, making it more difficult to replace their losses. In short, the fact that near-time extinctions struck almost exclusively large terrestrial mammals is in accord with the view that early hunters, not some climatic crisis, were mainly responsible.

Some have attempted to address the selectivity problem by arguing that climate change favored forage plants that were less nutritious for the large herbivores. As those herbivores died out, their predators and scavengers followed. A sudden, wrenching change in climate might hypothetically have reduced North American populations of trees and shrubs, replacing them with grasses and forbs. Such a massive disturbance might have reduced foraging opportunity for mastodons, which generally preferred woodland, riparian corridors, or forested habitat. At the same time, it would have favored mammoths, horses, pronghorn antelope, and other grazers. (Even mastodons, for that matter, ate some grasses.) Moreover, some of the known near-time changes in forage simply replaced one type of woodland with another, both suitable for the browsers of the time (Martin 1986, 124). And in general, large mammals find disturbed or successional vegetation, to be expected during postglacial warming, to be suitable forage (Martin 1990, 196).

Before the diets of large herbivores were well known, the extinctions were thought to have accompanied a shift from winter-rain grasses to summer-rain grasses resulting from the postglacial increase in summer rains (Kurtén and Anderson 1980). This theory is no longer accepted, however. Whether they require winter or summer precipitation does not compromise the forage value of the grasses; both types can provide good pasture. Mammoths, for example, ate both; they and other grass eaters were foraging on summer-rain plants in Arizona (Connin, Betancourt, and Quade 1998) and Florida (Koch, Hoppe, and Webb 1998) long before extinction struck. Summer-rain grasses evolved over five million years ago and are widely consumed by large herbivores, wild and domestic, in tropical and temperate latitudes north into Canada. Theoretically, then, climate change favoring either winter- or summer-rain grasses would be tolerable to large grazers.

Another indication that a shift in grasses does not explain the extinctions comes from Australia. Based on a variety of fossil eggshells of *Genyornis* (a flightless bird larger than an emu), Giff Miller and others (1999) date its extinction at 50,000 ± 5,000 years. They report that near that time, *Genyornis* in different parts of its range fed on either winter-rain grasses, shrubs, and trees or on summer-rain grasses.

Finally, it is difficult to imagine a single climate switch that would have affected all the large mammals in any given location simultaneously. Despite overlapping ranges, different species had different climatic preferences. For example, ground sloths were originally tropical, Harrington's goats boreal. A warmer climate should have been suitable for the sloths, a cooler one for the goats. Yet in the Grand Canyon they vanished at the same time. Climate-change proponents have offered various hypotheses to address such objections. Those with which I am familiar do not hold up to logical analysis. One, which argues for a shift from continental to maritime climates and back again, is well illustrated by a trio of related examples: the extirpations of tapirs *(Tapirus)* and vampire bats *(Desmodus)* in the United States and the simultaneous northward range shift of marmots.

Fossil records show that in the late Quaternary, marmots were found south of their present range, as at Rampart Cave, where they occurred at 1,500 feet, a very low elevation for marmot habitat even then. To climate-change proponents, this indicates that conditions were cool enough to bring marmots south from northern New Mexico and Utah, where they occur at present. In fact, fossil packrat midden records (Betancourt, Van Devender, and Martin 1990) do indicate an overall cooler condition during the last 40,000 years, until the early Holocene, 8,000 years ago.

On the other hand, the late-Quaternary disappearance of tapirs from Florida, California, Kansas, and Arizona (from the Sonoran lowlands to over 6,000 feet on the Colorado Plateau) is often attributed to a climate change in the opposite direction (E. Anderson 1984; Kurtén and Anderson 1980). Three species of tapir and vampire bats survive in the tropics of southern Mexico, through Central and into South America. Assuming that the present is the key to the past, many vertebrate paleontologists have surmised that for tapirs to have reached temperate latitudes in near time, the climate must have been warmer and the winters frost-free, as they are where many tapirs live now. The fossil presence of vampire bats in Arizona and California, north of their historic range (which covers tropical Mexico and points south), has also been taken to indicate late-Quaternary warming. Some of these bats were *Desmodus stockii*, an extinct species or population roughly 10 percent larger than the living *D. rotundus*.

To accommodate the movements of all three animals, paleontologists favoring climatic explanations have had to indulge in some contortions: the marmots came south because the summers became cooler. The bats moved north because the winters became warmer.

But neither vampire bats nor marmots live in coastal California now, in a climate free of the hot summers and cold winters of the continental interior. (In the late Quaternary vampire bats occurred on one of the Channel Islands off the coast near Santa Barbara, along with dwarf mammoths.) Moreover, plant fossils from packrat middens in Arizona and adjacent states do not suggest a more maritime climate in the late Quaternary. None of the packrat midden deposits at Rampart or nearby caves harbors plant assemblages typical of the California coast, such as soft chaparral. There was also no widespread northerly penetration of frost-sensitive tropical plants to match the range expansion of tapirs and vampire bats. Indeed, all the middens found in the Southwest (postglacial, late glacial, and full glacial) reflect arid conditions and a climate that while cooler was just as continental as it is now (Betancourt, Van Devender, and Martin 1990). When vampire bats haunted Rampart Cave and tapirs roamed the Colorado Plateau, they did not bring the rest of the tropics with them.

Some attribute the megafaunal extinctions to a shift from hypothetical warmer winters and cooler summers to the reverse (Hibbard 1958). The model shares the failings of other climate-change theories. It requires a unique climatic event, and one to which not only small but also large animals would have been vulnerable. In addition, it implies that large mammals were intolerant of continental as opposed to maritime climates. This is discordant with the present ranges of large mammals, which include some of the more continental climates known on Earth, such as those of the Himalayan and Mongolian plateaus.

The extinction of the large mammals, the likely source of the vampire bats' blood diet, probably did affect the vampire bats that once lived in Arizona and California. The miracle is that the vampire bats survived in the tropics. Very likely they multiplied and expanded their range with the historic introduction of livestock. As for the tapirs, it is more reasonable to model their late-Quaternary reduction in range as the result of confinement of refugial populations to tropical forest, where they were less vulnerable to new predators (people) than tapirs in more open country in temperate latitudes. A similar explanation may account for the fact that capybara *(Hydrochoerus)* and the spectacled bear *(Tremarctos)* are now limited to the tropics.

Finally, what about the marmots? Why did they not survive in Arizona and southern New Mexico? By the Holocene those regions had sizeable permanent human populations, and marmots are known to be popular prey historically. Historic range expansions northward of opossum, armadillos, and javelina, all attractive prey, may reflect similar change, with range expansion accompanying relaxation of hunting pressures since contact. I suspect that if introduced from the Rocky Mountains, marmots could be reestablished in Arizona.

The outstanding revolutions in the history of life—the global extinction of dinosaurs at the Cretaceous-Tertiary boundary roughly 65 million years ago and the extinction of many large mammals at the end of the Quaternary—were both long believed to reflect changes in global climate. Both can now be attributed to catastrophic perturbations unrelated, or only weakly related, to intrinsic climate change originating on our planet. John Alroy finds that "a consensus is forming that the end-Quaternary extinctions were caused largely, or possibly solely, by human impacts" (Alroy 1999). I totally agree.

Of course, the overkill model need not imply that climate change has never forced an extinction. In North America, for example, over the last 65 million years (the age of mammals after extinction of their Mesozoic prototypes), roughly 2,000 or more mammalian genera went extinct before our species arrived on the continent. Climate change is commonly invoked as the cause of these extinctions, and that may well be true in many cases. There are exceptions. Storrs Olson and David Wingate (2001) regard rise in sea level and associated drowning of the sizeable Bermuda platform as the cause of extinction of a newly described large flightless rail in the King Rail–Clapper Rail group, *Rallus recessus*. Their conclusion is supported by the absence of any historic accounts of the bird, and of any indication that prehistoric voyagers discovered and colonized the Bermuda platform.

Whatever exceptions those of us supporting the overkill model may be willing to make based on the field data, defenders of the climate school seem to feel they must warn the good people of Hamlin that Pied Pipers are dancing off with their intellectual children. For example, Don Grayson (2001) insists that those who study late-Quaternary extinctions most intimately, namely his generation of vertebrate paleontologists, strongly support a climatic as opposed to a cultural or some other explanation. Grayson's earlier objections intrigued one of his colleagues, paleontologist Peter Ward. In his book *The Call of Distant Mammoths,* Ward quotes Grayson as follows:

The results of that search [for reliable radiometric age dates on the extinct mammals] strongly suggest that overkill could not have been the force that Martin has claimed. The differential appearance of kill sites (only proboscideans, and within the proboscideans, almost only mammoth) and the strong hints that many of the taxa involved may have been on their way to extinction, if not already gone, by 12,000 years ago imply a far lesser human role in the extinction than the overkill model allows. The climatic models account not only for the extinctions, but for the histories of smaller mammals during the Pleistocene. With greater explanatory power, most scientists studying the extinctions issue accept climatic, not overkill, accounts, while recognizing that far more precision is needed in these accounts. This does not mean that people played no role in causing the extinctions. A multivariate explanation may yet provide the best account of the extinctions. But no matter what the human role might have been, overkill was not the prime cause of the extinctions. That cause rather clearly lies in the massive climate change that marks the end of the Pleistocene. (Ward 1997, 161)

Ward then adds:

In the mid-1990s, I was struck by the almost eerie similarity between Grayson's arguments [against overkill] and those being leveled against the impact theory for the disappearance of the dinosaurs. The proponents of both arguments believe that the victims—the Quaternary megamammals and the Late Cretaceous dinosaurs—were dwindling in diversity and abundance well before their extinction. Both assume that there would be a "bone bed" or that there would be more kill sites if the sudden, catastrophic explanation were correct. Both cite last occurrence dates (the time when the last known individual of any species occurs in the fossil record) for the victims as being well before their supposed final extinction. And both imply that they are correct . . . because those best acquainted with the facts agree it was not catastrophic. (Ward 1997, 161–162)

I am delighted with Ward's insight. Certainly there is a much greater opportunity to test for decline in range or numbers of near-time megafauna in America and Australia just prior to widespread human invasion than there is in the case of late-Cretaceous dinosaurs prior to the extraterrestrial impact that left the Chicxulub crater. Yet no reduction in the deposition of ground sloth dung or other common fossils of extinct species is evident. Mammoths, for example, apparently saw no decline in number of animals, number of taxa, or range right up until they went extinct (Martin and Steadman 1999, 42). The dates on the fossils of Harrington's goats also do not suggest a declining population (Mead, Martin, and others 1986). Early returns on dwarf ground sloth extinction in the Greater Antilles, a crucial test of the climatic model which predicts synchronicity, indicate survival of the dwarf ground sloths for thousands of years after extinction of their continental relatives. A simple climatic explanation of ground sloth extinction will not account for the extinction chronology.

In an exchange between Grayson and Dave Meltzer (2002, 2003) and Stuart Fiedel and Gary Haynes (2003), Grayson and Meltzer decry the scarcity of archaeological deposits associated with mammoths or other creatures from the more than 30 genera of large animals that vanished close to the time of human arrival. Fiedel and Haynes, on the other hand, think "there is far more support for overkill than for climate change as the principal cause of the extinctions." These four archaeologists do agree that the total number of unambiguous associations of human interactions with now-extinct mammals is represented by 14 proboscidean kill sites. I am glad to see that Grayson and Meltzer accept human impact as the cause of thousands of flightless bird and sea bird extinctions on oceanic islands. Only 40 years ago virtually no one, with the exception of Charles Fleming, attributed moa extinction to humans.

TEN

RESTORATION

A day of darkness and of gloominess, a day of clouds and
of thick darkness, as the morning spread upon the mountains:
a great people and a strong; there hath not been ever the like,
neither shall be any more after it, even to the years of many
generations. A fire devoureth before them; and behind them a
/ flame burneth: the land is as the Garden of Eden before them,
and behind them a desolate wilderness; yea, and nothing
shall escape them.

Joel 2: 2-3

The give and take of public lectures has often revealed places where my
interpretations needed more thought. For example, when I gave a talk
at a Texas university some time ago, a skeptic in the audience surprised
me with a question I have since learned to expect: "If humans killed off
the mammoths, horses, and ground sloths, how did the buffalo survive?"
According to Ernest Thompson Seton, early in the nineteenth century,
before market hunting began, North America harbored 60 million bison. In the conservation community, some number between 30 and 60
million is sure to pop up whenever early bison are mentioned. How could
there have been so many bison if the prehistoric First Americans and their
descendants were such potent hunters? Why wasn't the bison extinct?

I pointed out that this question could be asked about whatever animal
happened to be the largest of the survivors. If bison had slipped into extinction, one could ask, why not the moose? And if neither bison nor moose
had survived, one could inquire how elk had managed to hang on. Soon
we would have to frame the question around rabbits and packrats. "Be-

sides," I went on, "of several fossil taxa of American bison, all but the smallest became extinct. And after the main extinction event, bison became scarce; their range shrank. At times they are hard to find in the fossil record. Some have claimed the animals were victims of a hot dry climate in the mid-Holocene known as the Altithermal. But we know from the fossil record that bison were widely hunted. In Montana, there are jumps above thick deposits of bones of late-prehistoric bison. For whatever reason, the genus pulled through, but at times it was a close call."

Mulling over this question later, I realized that the estimate of 60 million was something of a red herring. In the first place, it is not clear that it was ever more than an educated guess (Kay 1998, 2002). Seton had extrapolated from regions where bison were historically abundant to others thought suitable for the animals, even if few or none occurred there in the nineteenth century. The seminal reporters on big game along the Missouri and Columbia Rivers, Meriwether Lewis and William Clark, could have told Seton that projecting numbers of bison across all suitable habitat in historic times was a mistake.

I have been hooked on Lewis and Clark's natural history ever since acquiring an edition of their journals some years ago, while attending a paleoecological meeting at the Mammoth Site in Hot Springs, South Dakota. Ransacking the journals for information about bison in the Dakotas, I spotted one riveting entry. Heading home in late August 1806, the Corps of Discovery bivouacked on the Missouri River near the mouth of the White River, in what became South Dakota. Here, near the end of their journey, Clark estimated seeing 20,000 buffalo in one afternoon, a record number in his experience. He also noted the killing of two porcupines, an unusual event because porcupines were so highly valued for food and quills that hunting pressure eliminated them near villages. Clark noted, "I have observed that in the country between nations which are at war with each other the greatest numbers of wild animals are to be found." The expedition was camped in a war zone (Martin and Szuter 1999). Lewis and Clark also found large numbers of fearless bison, elk, and wolves in another such zone, along the uninhabited upper Missouri and the Yellowstone River in Montana. Tribes hostile to each other ranged the periphery of this area but did not settle there.

Anthropologists have long recognized the existence of buffer zones, war zones, or neutral zones between warring groups (Hickerson 1970). Recently wildlife managers and ecologists have begun to consider the relevance of these zones in their own disciplines, modifying Hickerson's theory to apply, for example, to deer living "in a 'no-wolves' land" between territories of adjoining packs (Mech 1977; Martin and Szuter 1999, 38). Apparently such zones have had a profound influence on large-animal aggregations in the historic period (Martin and Szuter 1999, 2002). William Clark may have been the first to understand how bison could be so abundant in such a zone (Kay 1994; Martin and Szuter 1999). He also observed that they were fearless, another consequence of their separation from humans (see Jared Diamond's account of his experience with a tree kangaroo on an isolated mountain in New Guinea [1997, chapter 7]). The demilitarized zone (DMZ) between North and South Korea, mined and strictly off-limits to humans, and therefore safe for huntable wildlife, represents a modern war zone. To escape hunters during migration, two species of cranes seek out the DMZ as a refuge (Higuchi and others 1996).

More broadly, where human populations are denser, wildlife populations are usually smaller. This is another reason that European explorers of North America found themselves alternately in regions of scarcity and abundance. Only in uninhabited regions, such as along the Canoe River in British Columbia, did the early fur traders find abundant moose and beaver. In contrast, along the Columbia River near the Horse Heaven Hills, where salmon ran and edible wild plants were abundant, the river and its tributaries sustained large numbers of Native Americans. The Columbia was therefore a place where local populations of preferred prey such as beaver, bison, deer, elk, and pronghorn were few or absent. This was not because the habitat could not support numbers of these animals but because it did support a relatively large human population, ready and willing to hunt wild game as the opportunity arose. Metapopulation ecology (Pulliam 1988) treats the dynamics of populations within a large area, especially their fluctuations according to differences in habitat productivity or predation. With rare exceptions, early travelers, occupied with survival, overlooked this dramatic interplay.

In short, bison were probably never, at least since the arrival of the First Americans, as common near settlements as they were in uninhabited lands. Moreover, Seton's extrapolations appear to be based on the assumption that bison were in decline in the early 1800s. In fact, they were on the rise because Native American populations were declining due to exposure to European diseases. There is good reason to believe that human impact on the ecosystems of the North American West was ebbing when Lewis and Clark made their journey (Boyd 1999; Diamond 1997; Kay and Simmons 2002; Reff 1991). Old World contact beginning 500 years ago reduced Native American populations by as much as 95 percent; that is, only one person in 20 survived contact (Dobyns 1983, 1993). This was caused largely by the introduction of diseases such as smallpox and measles, but there was also warfare, exacerbated by the uneven availability of guns, trade goods, spirits, horses, and new religions, all of which increased the intensity of intertribal warfare. Even if the depopulation were much less severe, let us say 50 percent, it would have considerably diminished human predation pressure on big game.

Historic writings reflect this relaxation of Native American influence on ecosystems. In them we read of remarkably large numbers of game, including wild horses (introduced by the Spanish) and their predators and scavengers, including grizzly bears and wolves, enjoying trophic opportunities not seen since pre-Clovis times. When reported by Lewis and Clark, three centuries after the crisis of contact, bison were thriving as they had not for thousands of years, if ever.

Therefore, the question is not really "How could there have been 60 million bison in the early 1800s?" Rather, it is "How did *any* big game manage to survive intense hunting by early Americans?" The answer most likely varies with the animal and its behavior. Polar bears and grizzly bears, for example, are dangerous prey; the hunter all too easily becomes the hunted. Shortly before calving, caribou shed their predators on rapid, long-distance migrations north to the empty tundra of the subarctic. Bison move unpredictably across the vastness of the Great Plains, with no fixed migration route. When hunted, elk slip into dense cover, while mountain goats and mountain sheep retreat into rough country. Moose rely on a keen sense of hearing or smell to escape predators. Pronghorn race away at high speed. Deer reproduce rapidly and thrive in disturbed habitats (recently expanded to include the suburbs).

Of the various conclusions one might draw from all of this, the foremost is that we often identify as "wild" conditions those that are in fact heavily influenced by humans. In appraising ecosystems, both ecologists and the general public may overlook, or leave to the anthropologists, or simply take for granted the one mammal of overriding importance—*Homo sapiens*. Charles Kay (1998) has designated us the "ultimate keystone species." (Ecologists define keystone species as organisms that profoundly influence energy flow or habitat carrying capacity.)

At least until recently, many historians, conservationists, and ecologists have accepted historical documents such as those by Champlain, Coronado, or Lewis and Clark as reflecting the New World when it must have been "wild," "pristine," and "primeval" (see, e.g., Bakeless 1961). By definition, only Europeans could significantly influence "nature," which was essentially viewed as including native people. Similarly, in longing for a "last entire earth," Thoreau and others of his time had in mind New England before the Pilgrims, when, Longfellow poetically pronounced, murmuring pines and hemlock made up the forest primeval. Hemlock qualifies but pines are suspect. Recently ecologists have come to accept the significant role of native people in changing the land and its fauna before European contact (Kay 1994, 1998; Martin and Szuter 1999). Despite their value, historic records do not inform us about an America (or any other prehistorically colonized land) free of human impact. Nor do they inform us of the nature of the ecosystem when native people were at the peak of their powers, before the deadly epidemics of contact (for the Pacific Northwest, see Boyd 1999).

The view that preliterate societies made no difference began to shift as fossils of the Neolithic became known in the Mediterranean, a land profoundly altered long before Homeric times. It shifted again with discovery of the Aztec, Mayan, and Inca civilizations in America, to name a few, along with realization of the significance of the ancient African city of Great Zimbabwe. In New Zealand, palynologists (those who study fossil and modern pollen and spores) discovered that the open, patchy forest recorded by the first English explorer, Captain James Cook, and long thought to be primeval, was not. The pollen record showed that prior to the arrival of the first settlers from Polynesia, much of New Zealand would have been a closed forest (Anderson and McGlone 1992). With fire, prehistoric people opened the forest. Extinction of the moas soon followed.

Similarly, as discussed in chapter 6, the fossil record rarely supports the common assumption of earlier zoogeographers, including Darwin and Wallace, that whatever they found on any previously unstudied Pacific island represented "nature in the raw." In an unwitting form of racism, zoogeographers characterized as natural the fauna—native people included—that they encountered on islands and archipelagos. They assumed that only European voyagers could overhunt native species or introduce lethal aliens. In fact, prehistoric settlement had radically altered the fauna of these islands (Olson and Wetmore 1976; Steadman 1995; Steadman and Martin 2003).

It will come as no surprise that I define the "last entire earth" differently than did Thoreau. Prehistorians find that any given land begins to lose its wildness not when the first Europeans arrive, but when the very first humans do. In the Americas true wilderness was more than 10,000 years gone by the time Columbus reached our shores. It disappeared with the megafauna, whose calls gave voice to the forests and prairies.

This perspective is hardly the norm. Many of us were raised on the legend that what Lewis and Clark saw was Wild America, America the Beautiful. And in truth, America is not only the beautiful but the opulent. Vastly rich in resources, endowed with highly productive soils, enjoying for the most part a temperate climate, possessed of ample waterways and ports with protected access to the oceans, blessed with many natural areas displaying a rich assemblage of plants and animals, and inhabited by a diverse population of boundless energy, the Americas in general and the United States in particular are viewed with envy by many people around the world. Indeed, most Americans view their opportunities as boundless and their heritage as unique, a source of great optimism.

It can come as a shock to learn that in at least one respect this heritage is in fact woefully impoverished. A great many large animals, gifts of the evolutionary gods, were destroyed before anyone drew their images on bone or stone or on the walls of American caves. The near-time extinctions deprived North America of two-thirds and South America of three-fourths of their native large mammals. It was the remaining third that so impressed Lewis and Clark. The survivors are highly valued, but they fall far short of defining the natural fauna of the hemisphere. As far as large animals are concerned, America the Beautiful is now America the Blighted.

If we could travel back to near time, we would easily spot the sheer variety of unfamiliar large mammals distinguishing truly wild America. Perhaps less obvious, but no less important, would be differences in the faunal niches being filled, in the ranges occupied by various animals that still exist today, and in the impacts of all the animals on their surroundings. These more subtle observations would also have profound effects on our definitions of "nature." Though we cannot literally travel back in time, the comparable observations we can make based on the fossil record should open our eyes to possibilities.

For example, managers may assume that the species of plants and animals that a habitat has been known to support in historic time are a fair representation of how things were in the ice ages. The accidental or deliberate introduction of alien species is therefore viewed with alarm. Aren't aliens destructive? Reality is much more complex. From fossils we know that the grasses, forbs, shrubs, and trees of the Americas coevolved with a much greater variety of large herbivores than exists today (for California, see Edwards 1992). The absence of bison in California historically may well account for the nature of California grassland as reported in early documents (Bock and Bock 2000, 38). But that does not mean the native grasses were never subjected to heavy grazing either by bison or by the native horses, mammoths, and other megaherbivores found in abundance in the Pleistocene and older fossil faunas of California and the West. The grasslands and savannas of America coevolved with many species of large herbivores and presumably with heavy herbivory.

Heavy herbivory in ancient times may account for the fact that in arid regions of the southwestern United States and northern Mexico many low trees, shrubs, and sub-shrubs are armed with thorns, oils, terpenes, and/or tannins designed to repel herbivores or discourage excessive consumption by them. Even some of the trees in dry tropical forest communities are armed with thorns. Examples in Africa, southern Mexico, and Central America include the young *Ceiba,* whose trunks have thorns, and the spiny acacias, which have large hollow thorns colonized by ants ready to assault intruders. The spiny nopaleras studied by Dan Janzen and the deciduous tropical forests in Costa Rica and other dry parts of Central America and Mexico feature sweet fruits or pericarps that are ingested by large animals, which become unwitting agents of seed dispersal (Barlow 2000). Ethnobotanist Gary Nabhan tells me that Native Americans may have unknowingly substituted for the large extinct quadrupeds in dispersing devil's claw and squash seeds. Subsequently, horses and cattle helped in seed dispersal.

These rarely acknowledged changes illustrate the hazard of accepting the current extinction-pruned large mammal fauna of the Americas as the "normal" evolutionary assemblage (Janzen and Martin 1982; Barlow 2000). The disappearance of the megafauna opened ecological opportunities for many kinds of large animals, including (but not inherently limited to) those found historically. The fact that "new" animals took advantage of those opportunities makes the historic record an even more deceptive guide to any true state of nature. Big-game ecologist John Teer puts matters this way: "Some people say that success for introductions of foreign animals became a foregone conclusion when the original Quaternary fauna was lost, perhaps because of overhunting by early humans" (Mungall and Sheffield 1994, ix).

For instance, take the question of the "natural" range of bison and the related issue of what lands are naturally suited for grazing. In the time of Lewis and Clark, native people and their horses, the latter estimated to number in the hundreds of thousands, occupied the Columbia Plateau in Washington. In the absence of people, I believe that the Great Basin sagebrush-grasslands would have swarmed with bison. The success of cattle in much of the West confirms that bison could have thrived in areas outside their historic range (Martin and Szuter 1999). The historic absence of bison in southwestern New Mexico and Arizona, for example, may reflect the marginal carrying capacity of desert grassland and the hunting skills of Native Americans living in this region, alternately raising crops *(Zea)*, gathering wild plants, hunting wildlife, and, when they could, hunting bison (Speth 1983). It need not reflect a range unsuited for bovids.

Very likely before extinction the near-time fauna harvested energy from plant communities in different ways than did the historic fauna. In the process, the mastodon, long-nosed peccary, stag moose, and Jefferson's ground sloth dispersed fruits and seeds of temperate forest trees more effectively than white-tailed deer, the largest surviving native herbivore in much of the East. Thus, the current pattern of change in plant communities should be quite different from those common in the time of heavy usage by the Quaternary megafauna. Currently ungrazed or unbrowsed floodplains need not represent a natural stable state. American floodplains may have many potential stable states. The overgrown willows and giant senescent cottonwoods often seen in "protected" floodplains would likely not prevail in riparian habitat open to proboscideans, equids, bovids, and cervids, the dominant species of the stable state preceding megafaunal extinctions.

We should also appreciate the dynamic nature of Cenozoic ecosystems. For instance, it was not until late in the Cenozoic, a scant 200,000 years ago and tens of millions of years after the radiation in the Western Hemisphere of numerous equids and camelids, that the family Bovidae, to which bison, cattle, and eland belong, finally entered the New World. The Cenozoic history of New World mammals was endlessly dynamic and changeable (Flannery 2001). Understanding the Cenozoic evolution of mammals is vitally important in gaining perspective on how one might design with nature and what might happen in the future. For example, in the absence of our species, the next drop in sea level would foster further exchange with Eurasia across the Bering Land Bridge.

Despite all we have learned about the fossil record, until recently the fauna recorded over the last five centuries has been uncritically accepted as an American baseline; it is rarely viewed as a proposition or considered a work in progress, to be subjected to penetrating analysis. We yearn for "a home where the buffalo roam, where the deer and the antelope play" (Brewster Higley, "Home on the Range," 1873; see Geist 1996). Native glyptodonts, ground sloths, and proboscideans go unmentioned.

Those who dream of and work toward preserving and restoring the

American wilderness generally ignore the fossil record. Conservationists hammer home the message that we live on a plundered planet with extinctions accelerating worldwide. So far so good, but this analysis does not look back far enough in time. Some feel that only those species historically known in North or South America belong here, and only in their historic ranges. (Problems may arise when animals themselves expand those ranges, as in the case of elk and bison entering Grand Canyon National Park.) None of this "speciesism" considers the influence exerted by the ultimate keystone species, *Homo sapiens*. The classic vision of a restored prairie, for example, involves herds of bison, elk, and pronghorn followed by wolves, amid prairie dogs and burrowing owls. Such dreams are certainly preferable to monocultures of range cattle, but they fall short of nature in evolutionary time. The Cenozoic has yet to be given its day in court.

Not only the philosophy but the practical efforts of conservationists have focused on preserving or restoring the "wild" America known to us from historic time. It is considered vital to protect caribou in the Arctic wilderness; to cater to cougar, mountain sheep, and mule deer in the Grand Canyon; to keep bison, equids, and mountain goats off lands where they do not "belong." All this is believed especially important in our national parks, selected for their allegedly undisturbed character and viewed as benchmarks for land management, places where natural conditions deserve to reign supreme.

Accordingly, following release of the Leopold Report (named for Starker Leopold, chair of the committee that prepared it), the National Park Service in the 1960s adopted a policy of maintaining national parks in the condition they would have been in 500 years ago at European discovery (Leopold and others 1963). Later arrivals, such as cattle and horses, would not be tolerated in the parks if they could be eliminated. Of course, a great many alien species, such as English Sparrows, carp, dandelion, and red brome and hundreds of other species of nonnative grasses, cannot be purged or even identified easily. They persist in national parks by default.

In 1982, for instance, rangers in Colorado National Monument got rid of their herd of American bison, introduced by the town of Grand Junction 50 years earlier. Fenced within the monument, the bison grazed it heavily. Managers could not sell or destroy excess animals, and, under pressure from ecologists, the NPS decided that all the bison had to go. This might not have happened if the animals had occupied western Colorado around Grand Junction in the 1800s.

Bison seem to be especially vulnerable to the view that historic ranges should be cast in concrete. Though bison thrive when given access to most rangelands, it would not be considered "natural" to let them live in Arizona, California, western Colorado, Idaho, Nevada, western New Mexico, West Texas, most of Oregon, western Utah, and Washington. The reason? They were not found in those areas historically (see range maps in Graham and Lundelius 1994). This view often accompanies an antigrazing philosophy: in the historic absence of bison, exclude livestock.

In another example of the struggle to do right by nature, the retention of long-established and possibly natural Rocky Mountain goat populations in the Olympic Mountains on the Olympic Peninsula in Washington occasioned a great outcry. Some conservationists argued that the animals were not there naturally and that they severely damaged endemic alpine plant communities. A thorough treatment of the controversy (Lyman 1998) reveals the heavy hand of some activists in approaching historical, biogeographic, and paleontological uncertainties. On their travels along the Columbia River, Lewis and Clark saw no live goats. They were shown blankets made of mountain goat hair. Very likely the goats would have been preferred prey, heavily hunted and, because of their restricted range, vulnerable to prehistoric extirpation. Perhaps the Olympics were an outpost. Does this make their presence there today blatantly "unnatural"?

Some ecologists think range grasses have lost their adaptability to grazing in the 10,000 years since the extinctions (Belsky 1992). Designing an experiment to test such a proposition is not as difficult as one might imagine. It appears that the experiment has already been run for us. Summer-rain range grasses, many species in the genera *Aristida, Bouteloua, Muhlenbergia, Sporobolus*, and others, flourish on both sides of the boundary between Arizona, New Mexico, and West Texas and Sonora and Chihuahua. Satellite imagery indicates heavier grazing on the Mexican side now. Nevertheless, summer-rain grasses are much more speciose in Mexico than in the United States.

In the 1970s Grand Canyon National Park came under pressure to get rid of its wild burros. For many years, whenever it was felt that burros had grown too numerous, Grand Canyon's park rangers shot them, killing a total of 2,860 between 1924 and 1969 (Carothers, Stitt, and Johnson 1976). In the 1970s, however, GCNP managers learned that shooting would no longer serve to control the burros and decided that they must be removed. Various reasons were cited for this. All of them ignored the fact that New World vegetation had evolved in the presence of herbivory by equids.

One of the reaons given was the burros' alleged potential for uncontrolled expansion. Around this time, as part of a research project in the western part of their range, zoologist Steve Carothers shot and autopsied 150 animals. That virtually eliminated the Lower Canyon herd. In his autopsies Carothers found no predators or parasites that might control the burros' numbers, and the reproductive organs he examined supported other evidence that burros could increase at 11 to 17 percent annually. Although it was coming into fashion in the management of elk in Yellowstone, natural regulation (in the form of limited forage, winter mortality, and birth rates that decline as nutrition becomes poorer) was not considered an option for burros in GCNP. Some felt resources were ample and the burros would soon find themselves with standing room only. However, historical accounts indicated that the Lower Canyon herd had lived there for roughly a century and, unlike herds elsewhere in the park, had not been controlled by shooting. It therefore seemed obvious, to me at least, that the Lower Canyon burros must have been self regulating. In one hundred years, at an average annual rate of increase of 10 percent, an initial group of 10 animals could have grown to over 10,000. Nothing like that had happened. The Lower Canyon herd at least should have been spared.

Burros were alleged to be severely damaging native plants, including coach-whip, mesquite, and grasses, in the 25 percent of GCNP that they allegedly occupied. I am satisfied that actual percentage was much less, based on what I saw from a helicopter ride over the terrain available to the Lower Canyon herd. (Much of the Grand Canyon is too steep for burros.) In 1979, with travel support from Joan Blue and the American Horse Protective Association, I hired Wayne Learn to fly geologist-photographer Pete Kresan with me over the region that the Lower Canyon herd had once occupied. We were in search of burro damage. Well above the Colorado River we spotted a faint burro trail, much less obvious than many trails used by hikers in the park.

We landed in the mouth of Two Hundred and Nine Mile Canyon, where destruction of mesquite and other damage had been claimed. Across the river, in Granite Park, the NPS field team had reported no burros and no damage to mesquite. What we saw on the ground did not convince me that burro impacts were alarming or intolerable. Had enough research been conducted? Burros were charged with hammering the decorative, common, and widespread ocotillo, a specialty of southwestern desertscrub also known as coach-whip *(Fouquieria splendens)*. Within burro range across the river to the west of Granite Park, I found some evidence of bark stripping of coach-whips, adjacent to numerous untouched coach-whips growing on steep slopes and ledges, out of reach of burros. In the Grand Canyon there are many such ledges, where coach-whips and other palatable plants are out of reach of burros and even of mountain sheep.

I had been stunned to read that burros savaged mesquite, a shapely low tree remarkably resistant to destructive herbivory by either domestic or wild animals. Mesquite sprouts vigorously if axed, smashed, bulldozed, or, in an occasional extreme winter, frozen to the ground. The new growth is heavily armed with stout thorns. In northern Mexico heavy use of the land by livestock, including cattle, horses, and burros, goes hand in hand with an abundance of mesquite. Ranchers would be delighted if burros or any other animals *could* control or eliminate mesquite, which is notorious for its invasion of rangelands in the last century, to the detriment of pasture grass (see repeat photographs in Hastings and Turner 1965; Turner and others 2003). In fact, like other domestic animals, burros help spread mesquite. Substituting for extinct animals of the Quaternary, they devour the sugary bean pods, dropped to entice large herbivores at the season just before summer rains, when other provender is scarce. If park managers wanted more mesquite, they would want to keep some burros to help distribute its seeds.

Driving through the Hualapai Reservation from Peach Springs down the spectacular Hurricane Fault to Diamond Creek, a steady descent of over 4,000 feet that is a short course in biotic change with elevation, I often see burro droppings and occasionally the animals themselves. I have not looked hard, but a casual inspection reveals no severe damage to coach-whip, mesquite, or other native shrubs. To be sure, I understand that at times the Hualapai harvest their burros.

Like other equids, burros graze. However, the main grass identified in their dung in the Grand Canyon was not a native but a *Bromus*, very likely the introduced red brome, *Bromus rubens* (Hansen and Martin 1973). Pete Kresan and I found much less red brome on the west side of the river under mesquites allegedly killed by burros than under unmolested mesquites across the river. Were the burros helping to control an alien plant? Picking sides was not as simple as the anti-burro faction claimed.

The bottom line, as far as I could tell, was not that the burros were causing great damage to the park, but simply that they were not "native" to it. It all boiled down to the Leopold Committee's decision on wilderness. If burros and other equids did not graze in wild America 500 years ago, it did not matter where their ancestors had lived for the pre-

vious 50 million years. The burros would simply have to go. Private organizations such as the Fund for Animals offered to help pay to round
them up and remove them from the canyon by whatever method worked.
If all else failed, however, the burros would be shot.

On October 14, 1977, GCNP resource managers scheduled a burro
workshop. Despite our heretical views, one of the organizers, Jim Walters, who had prepared the NPS impact statement and was not in favor
of the burros, gamely found spots on the program for Ken Cole, Geof
Spaulding, and me. Hoping to tempt the audience into taking the long
view, I brought along 50 feet of yellow nylon rope, with each foot representing a million years. Black electrician's tape covered the last half
inch.

When my turn came, Ken ran the rope through part of the auditorium.
I waved the end with the black tip and explained that if the rope represented the 50-million-year evolution of equids in the New World (including, of course, their coevolution with native plants), that last half inch
represented the interval in which they had been missing from North America. I ventured an opinion that the 10,000-year absence of equids from
GCNP had been unintentionally remedied a century ago by their return,
in the form of miners' burros, to a small part of the park. The burros might
well occupy the same ecological niche as the small extinct equids known
from Stanton's and other caves. Were "defenders of the truth" ready to
relax their grip on the deeply entrenched view that whatever was found
historically represented the only truly "natural" biota of America?

Finally, if park managers really wanted to reduce impact on the vegetation and soils of the canyon, they could close down heavy mule traffic
on the trails to Phantom Ranch. A tramway would deliver far more visitors to the inner gorge with far less impact on the park (I thought of
the U.S. Guano Corporation's operation). And if riding mules and pack
mules were retained because they were traditional, well, so were the wild
burros. Miners had released them in a few parts of the canyon, where
they were still confined, long before steps were taken to establish a national park.

We could tell, however, that our efforts would be unavailing. NPS
biologists rarely get a chance to improve, as they see it, on policy, and they
were not ready to shift gears on this one. The idea of a refuge for burros in any national park or monument got a thumbs-down. If the cost
of keeping the parks "natural" was burro extirpation, so be it. Later I
learned that an unregulated population of tens of thousands of elk severely suppressed serviceberries in Yellowstone National Park, depleting

habitat for grizzly bears and beaver (Wagner and Kay 1993; Kay 1998). NPS biologists regarded this as an acceptable case of natural regulation (or lack thereof). In contrast, even if burros did self regulate in GCNP, they could not earn citizenship. They were aliens and therefore bad guys, and that was that.

Given the recent efforts to reestablish condors in the Grand Canyon region, the U.S. Fish and Wildlife Service might wish it had more large animals whose mortality would provide food for the scavenging birds. Fortunately, wild burros and horses survive in that part of the Grand Canyon embracing the lands of the Hualapai and Havasupai Nations, which do not fall within the jurisdiction of the NPS. Recently Steve Carothers wrote that the Hualapai Nation is interested in negotiating with the Fish and Wildlife Service about turning some of its burros into fast food for condors (Carothers 1996). Perhaps these burros might someday also serve as prey for reintroduced wolves. In addition, there is much more to learn about plant-animal coevolution in the Mojave Desert thorn-scrub. With the Hualapais' approval, the burros can help. Am I wrong in thinking that they do not eat more than the bean pods of mesquite? Meanwhile, near the lower end of the Diamond Creek road picnickers can get acquainted with the biota of the Mojave Desert thornscrub, burro dung, and sometimes the burros themselves.

Few would deny that whatever their cause, American extinctions in near time impoverished the ecosystem. Should we now stand by passively in the face of efforts to throttle back reintroduced equids, after 50 million years of remarkable success of the group? These proxies for their lost relatives are, I submit, better than nothing, especially if campers at Diamond Creek in the Grand Canyon are lucky enough to be roused from sleep by the hair-raising bray of a proxy echoing off canyon walls at midnight.

In the early days in Yellowstone, for "the good of nature," rangers exterminated wolves and killed other predators that might prey on deer and elk. I imagine those rangers felt virtuous and glowed with the adrenaline rush that hunters experience after a kill. The cessation of this campaign shows that park policies and public attitudes are reversible (Martin 1996). Now that large predators are appreciated as essential ingredients in ecosystems, wolf reintroduction is well under way in certain Western states. I would guess that those releasing wolves feel virtuous, and I would not be surprised if they experience an adrenaline rush of their own when the captive wolves emerge from their transport cages. In the fullness of time, will the NPS and its stewards reconsider its exclusion of wild burros from parts of the Grand Canyon?

"Natural" changes in animal distribution also complicate conservation issues. In recent years, for example, elk have begun to invade GCNP—and not the native Merriam's elk, which went into eclipse over a century ago, but introduced Nelson's elk, *Cervus elaphus nelsoni* (Hoffmeister 1986). Recently javelina (collared peccaries) also began approaching GCNP, from Cameron, Arizona. Although javelina are commonly viewed as an emblem of the Sonoran Desert, the zooarchaeologists I have contacted report not finding their bones with those of deer and mountain sheep in prehistoric refuse. Apparently, javelina survived the late-Quaternary extinctions somewhere in Mexico and moved north only in the last few hundred years, replacing larger peccaries *(Mylohyus, Platygonus)* that were native and had gone extinct thousands of years earlier. Therefore, as well as they fit in, javelina are hardly more native to GCNP than the wild burros. Most recently, bison have invaded the North Rim forests, breaking out of a droughty pasture in the adjacent House Rock Valley to seek lush grasses at higher elevations. The bison herd is maintained by Arizona's Department of Game and Fish; the animals are shot by license. Here we go again.

To sum up, many conservationists have defined the large mammals of written history as representing nature's intentions for the New World. This "Columbian curtain" locks us into a few hundred years of ecological time, blocking out not only the largest and most representative animals of the continent, but those with the longest tenure. Of the large mammals present historically, only bison is represented among the more common fossils of the late Quaternary. It shares billing there with an abundance of equids, camelids, and especially proboscideans (Graham and Lundelius 1994), all of which were here much, much longer.

Interestingly, various human acts in recent years have—intentionally or not—moved North America a bit closer to pre-Columbian, pre-Holocene wilderness. These unheralded experiments in near time resurrection have fallen into two general categories. Some have reintroduced species that are taxonomically identical or closely related to those lost. Others, recognizing the dynamic nature of ecological guilds, have introduced species that might perform the same ecological functions as the missing ones, even if the proxies are not taxonomically related.

In the first category, one unplanned restoration has been under way for centuries. When the Spanish arrived in America in the 1500s, they brought domestic horses with them. These animals were much closer genetically to one of the extinct American horses *(Equus caballus)* than some have realized. Widely domesticated in Eurasia, *Equus caballus* had multiple points of origin, unlike many other domestic animals (Vilà and others 2001). In less than 200 years after the Spanish arrived, wild horses spread from the Mexican Plateau into Canada and throughout the pampas in Argentina. Hundreds of thousands, if not millions, roamed the grasslands of both North and South America. Along the Columbia River west of the Rockies, Lewis and Clark found tens of thousands incorporated loosely into the economy and even the religion of the native people of the Northwest (Martin and Szuter 1999). Since neither the Spanish conquerors nor Lewis and Clark knew that horses had evolved for tens of millions of years in North America, vanishing only around the time of the last mammoths and mastodons, they could not know what they were witnessing: the extraordinary return of a species to its phylogenetic homeland. In 1971 Congress passed the Wild Free-Roaming Horses and Burros Act, making the animals the property of the United States and guaranteeing them a future.* It is now a crime for anyone but the government to kill, disturb, or capture wild, free-roaming horses and burros.

In recent years the Turner Foundation took what some felt was a very bold step. It purchased large ranches in the western United States and substituted buffalo *(Bison bison)* for cattle *(Bos taurus)*. When possible, the Turner ranches restored prairie dogs, black-footed ferrets, and Aplomado Falcons to ranges they had once occupied. Though all of these species were present in western North America historically, in some quarters these efforts are not viewed favorably. Once again, the logic is that bison are not known to have occupied these New Mexico lands in the time of Coronado or any other writer, and that there must be a reason for this. Maybe the land naturally belongs to pronghorn, to mule deer, and, where rough enough, to mountain sheep. Maybe other animals would damage the plants and the entire ecosystem. On the other hand, bison are increasingly popular as a beef animal, and near Truth or Consequences, New Mexico, over 1,000 bison, as well as prairie dogs and mountain sheep, recently replaced cattle on one of the Turner properties, the 600-square-mile Armendaris Ranch (where they were unknown historically).

The most intriguing restoration project I have seen in a lifetime of looking is to be found at the Canyon Colorado Equid Sanctuary in the High Plains of northern New Mexico. Here representatives of the living equids share lebensraum in blue grama–juniper grassland, on the continent where

---

* *Wild, Free-Roaming Horses and Burros Act, U.S. Statutes at Large* 85 (1971):649–650, codified at *U.S. Code* 16, secs. 1331–1340.

their ancestors evolved and from which they crossed the Bering Bridge in the Pliocene to colonize central Asia and Africa. The taxa present include Przewalski's horse; two species of zebra; ponies; quarter horse–size representatives of *Equus caballus;* kulan; and various taxa of African asses, including burros, *Equus asinus.* They differ importantly from one another in color, size, and behavior. In fact, they display the evolutionary diversity found for millions of years in the American West. On this ranch the High Plains habitat reveals its true potential.

In a less far-ranging example of restarting evolution of missing megafauna, Nelson's elk were brought into the southern Rockies after endemic Merriam's elk were extirpated in Arizona. Few managers had trouble with that, although the prehistoric and early historic elk populations were minuscule compared with those to be found in the West now (Truett 1996). Mortality of elk could add to the resource base for recently reintroduced California Condors and, potentially, reintroduced wolves.

Along with moose, mountain goats *(Oreamnos americanus),* close relatives of Harrington's extinct mountain goat, have been introduced into the San Juan Mountains in southwestern Colorado, where again there is neither fossil nor historic evidence of their earlier presence. The goat populations are thriving (Peek 2000). Rocky Mountain goats might also succeed as surrogates for Harrington's goat in northern Arizona, though the plants there are plateau, not alpine, species. The taxonomic separation between extinct and living goats is minimal, and the potential benefit to the reintroduced condors is an important consideration. According to Jim Peek (personal correspondence, February 24, 2004), the mountain goats do not require an alpine environment, just precipitous slopes in a rugged landscape. There they are safe from most predators, although ravens circling them tightly, as ravens are known to harrass humans, may induce a fall from vertigo.

An earlier experiment briefly reintroduced camelids to their ancestral home in the American West. Before the Civil War, the army tried using both dromedaries and Bactrian camels as beasts of burden in warfare against Indians. They transported water and supplies for military parties through the Southwest, establishing a major trade route through northern New Mexico and Arizona into California (roughly following what would become the popular Route 66). According to Lieutenant Edward Beale, who was in charge of the party (Stacey 1929), the camels ate the otherwise worthless weeds and other plants shunned by livestock, including creosote bush growing along the right of way in New Mexico. While initially successful, the project suffered a major political handicap: its sponsor, Secretary of War Jefferson Davis, soon found himself president of the Confederate states in the Civil War. The camels ended up outside Fort Tejon in California.

In a recent and more carefully researched effort at reintroduction, the Peregrine Fund has released captive-bred California Condors along the Vermilion Cliffs in Arizona, within 20 miles of Stanton's Cave, where many fossil condor bones have been found. Approved by the Fish and Wildlife Service under Interior Secretary Bruce Babbitt, this bold step began with the release of six condors in December 1996 (Snyder and Snyder 2000). Many public and private conservation groups supported the program. The birds soon began to range widely. They checked out inaccessible holes in the Redwall—a few had been ancient nesting sites for their kind—and began to use them, laying eggs and raising young. In the summer of 2000, the worldwide count of living California Condors was 171. By May 2003, it had reached 220, with as many as 20 soaring over the South Rim (Bob Audretsch email, May 28, 2003). Besides delighting park visitors with the beauty of their flight (Osborn 2002), the condors' restoration helps us overcome two blind spots. One fails to see the land as it was over 13,000 years ago; the other fails to see what it might be in the future. The Grand Canyon is once again the Valley of the Condors, as radiocarbon dates on fossil condor bones tell us it was in the late Pleistocene.

The second category of species introductions is potentially more controversial. Some of the extinct American fauna have no direct phylogenetic heirs or even, in some cases, close relatives. They may, however, have functional equivalents—guild members that can play comparable roles in the ecosystem. Attempts to introduce these guild members may raise more complex questions about management and louder cries of "alien." On the other hand, there are, to paraphrase the old cowboy ballad, not only "empty saddles" but also "empty niches by the old corral." And we have long been running these experiments in reverse, with livestock serving as proxies for extinct bison, equids, and camelids. Range managers should have insights of considerable interest on this front.

In 1963 I sent various conservationists a draft proposal for experimental African game ranching (then a new managerial approach) in the New World. My efforts ended up on the back burner, but I did publish an article in *Natural History* lamenting the lack of browsers in arid lands and noting the army's success with introduced camels in the 1850s (Martin 1969). Unexpectedly, I received a follow-up call from a reader named Julian Biddle.

If Cabots and Lodges are likely to be Bostonians, Biddles are Philadelphians. Julian was no exception, but he no longer lived in the City of Brotherly Love. In midcareer he had quit teaching Russian and Russian literature at the University of Illinois and purchased a 500-acre ranch 40 miles southwest of Tucson. But he had no intention of raising horses or cows. His move, he claimed, was inspired by what I had written about the potential for exotic large animals in the Southwest. What species, he asked, would I recommend?

For his limited acreage, laced with mesquite *(Prosopis velutina)* and catclaw acacia *(Acacia greggii)*, I suggested a small African browser, the gerenuk *(Litocranius)*, which will stand on its hind legs to browse low branches. Julian could find no breeding gerenuk for sale in the United States, however, and a strict quarantine law complicates introducing exotic animals from Africa. Fortunately, while searching for gerenuk, Julian fell in love with the common or lesser eland *(Taurotragus oryx)*. Eland are more than handsome—they are svelte. Their hooves click when they walk. If Herefords, one of the most popular breeds of range cattle, were as attractive, I think there would be much less objection to grazing on public lands. Behind the 8-foot fence required by the Arizona Fish and Game Department, Julian released Rufus and Sadie, acquired from the Catskill Game Farm in New York State. In time he added others. The high fence was prudent. Eland are great jumpers; Mungall and Sheffield (1994) say that one can leap over another practically from a standing start. A small group of University of Arizona faculty and students gravitated to Julian's operation. As far as we knew, eland were new to the Arizona range. How might they differ from cattle as herbivores? Were they in the same niche, or in one of their own?

Following the eland in his Jeep, Julian soon made some interesting observations. Though well supplied with commercial feed, his animals ate a variety of plants not especially favored by cattle. Undeterred by thorns, eland browsed on the summer green leaves of both catclaw acacia and mesquite. For his dissertation, a young Ethiopian graduate student in the School of Renewable Natural Resources of the University of Arizona, Ahmed Nasser Abdullahi, studied food habits of the eland using epidermal histology, Dick Hansen's technique of identifying plant tissues under the microscope. Ahmed kept two heifers in Julian's pasture as controls. During his brief study the dietary overlap between the eland and the heifers was only 15 to 32 percent. The heifers ate mainly grass. Eland took more forbs and browse. Thus, the species were mostly complementary, not competitive, in their feeding habits (Abdullahi 1980).

Eland already were introduced elsewhere in the Southwest. By 1988 there were 781 on Texas ranches (Mungall and Sheffield 1994). They did not require supplementary feeding.

For its part, the state of New Mexico has acquired African gemsbok *(Oryx gazella)*, greater kudu *(Tragelaphus strepsiceros)*, impala, Iranian ibex, African ibex, and aoudad sheep, initially acclimating them in a pasture near Red Rock on the Gila River in the southwestern part of the state (MacCarter 2000). The gemsbok, also known as fringe-eared oryx, have done particularly well and now thrive on the White Sands Missile Range and have entered Turner properties adjacent to public lands. According to range biologist Patrick Murrow (in MacCarter 2000), gemsbok have spread into "basins, foothills, grassy playas, rocky ravines . . . including pinyon juniper woodlands as high as 8,000 feet." They now number over 3,000. The introduction of Iranian ibex, confined to and hunted in the Florida Mountains in the boot heel of southwestern New Mexico, is also considered a success.

The Texan and New Mexican introductions of exotics are aimed at supplying private parties of hunters with unusual targets at a lucrative price. Both the gemsbok and the ibex are hunted. Some wealthy Texan ranchers are said to harbor black rhinoceros, the outcome of a Safari Club project initiated when poaching threatened black rhino in Africa.

Unknown in the American fossil record, members of the genus *Oryx,* including gemsbok, are African aliens, and most conservationists feel they should be denied New World citizenship. However, gemsbok flourish in desertscrub and arid grassland, land that is marginal alike for historically native large herbivores, such as deer and mountain sheep, and for livestock. Bill Huey (now retired from the New Mexico Department of Game and Fish, where he worked with introduced oryx, kudu, and other species) has told me that gemsbok are more active than cattle, not grazing very long in any one spot, and that their impact is minor compared with that of cattle (personal communication, September 2001). I suspect that the grazing niche now occupied by gemsbok would once have supported bison, equids, and llama-size camelids, all presumptive occupants of the same guild. It has been empty since camelid and equid extinction. Oryx are efficient herbivores in a land where large-animal herbivory suffered severely in near time.

In a specific example of the potentially positive ecological impacts of "non-native" proxies, in 2003 Ed Marston, then publisher of *High Country News,* reported that a New Mexico rancher plagued by an excess of nonnative tamarisk (salt cedar) had obtained good results by contracting with an owner of cashmere goats. She believes "that if the goats return to the ranch for a few weeks at a time over the next three to five years, the tamarisk will weaken and die." Then it is possible that lakeside grasses, willows, and cottonwoods will gradually come back.

Understanding our near-time losses suggests conservation and wildlife management approaches that go far beyond not only the modest, often unplanned experiments described above, but our most ambitious wilderness preservation efforts to date. Some activist conservation biologists are beginning to talk about what has long seemed impossible: moving beyond "restoration ecology" to the resurrection of the foraging behavior of animals now buried in the graveyards of near time.

UNEXPECTED RAMIFICATIONS
OF ECOLOGICAL CHANGE

Surprisingly, the megafaunal extinctions may have had significant consequences for the political history of the Americas.

According to Jared Diamond (1997), the absence of large and potentially domesticable mammals in the New World made a fatal difference when European invasion began. The extinction of native horses had eliminated any chance for New World people to domesticate their own riding stock. As a result, Moctezuma lacked cavalry to help fend off the Spanish invaders. In addition, the lack of domestic animals to match the sheep, pigs, chickens, ducks, and larger domestic species that lived in close proximity with their Old World owners reduced the number of endemic New World diseases. In the Old World, pathogens evolved in the mix of domestic animals and their human owners.

At first glance the scarcity of endemic diseases might seem to have been an asset for American Indians. It was not. In the ecological crisis of contact, America's native people had no immunity to the virulent pathogens brought over by the Europeans and their African slaves. Smallpox, plague, cholera, undulant fever, measles, and other viral and bacterial diseases had coevolved in the Old World with domestic animals and people. With the possible exception of the spirochetes of syphilis (and this is debated), Native Americans had no endemic pathogens that could in turn infect the newcomers.

The Spanish armies setting out to conquer America attributed their incredible success to God's will. Instead, the unseen force that aided them in battles in Peru, in the Valley of Mexico, and elsewhere was a truly insidious fifth column. Smallpox and other Old World diseases swept away large numbers of Native Americans—up to 95 percent, according to some anthropologists.

ELEVEN

RESURRECTION

The Past Is Future

The future is the largest of all possible subjects.

Bruce Sterling, *Tomorrow Now*

"Resurrection ecology" does not refer to cloning from ancient DNA but to restarting evolution of at least some of the lost lineages.* As intriguing as such a project may sound, however, why should we consider it, given the difficulty we face in restoring even a pre-Columbian "state of nature" in the Americas?

What I have learned about Quaternary life convinces me that although our desire for conservation and wilderness restoration is admirable, and our efforts have been noble, our present goals are historically shortsighted and far too tame. We are obsessively focused on protecting what we have and utterly unaware of what we have lost and therefore what we might restore. No American terrestrial habitat, from sea to shining sea, has been "natural" for some 10,000 years. Fought-over wilderness areas such as the Arctic National Wildlife Refuge, though vital habitat for remaining megafauna, are already depleted of other large species. Indeed, the only truly pristine faunal wilderness left in the world are the pinniped haulouts and penguin rookeries in Antarctica.

---

*Even with dramatic advances in our ability to replicate and analyze ancient DNA, including DNA preserved in the carcasses of frozen mammoths or the dry excreta or tissues of ground sloths from desert caves, there seems to be no realistic prospect of genetic resurrection.

This is no reason to relax our efforts to preserve and restore the American wild. To the contrary, our losses enhance the value of what remains. Nevertheless, we are half blind if we behold the Grand Canyon without visions of its ground sloths, Harrington's mountain goats, California Condors, teratorns, and, on the plateaus embracing it, Columbian mammoths along with extinct species of bison, tapirs, camelids, and horses (Nelson 1990). As we have seen, some efforts to restore the "wild America" of historic time have actually taken us even further from the wilder America of near time.

If we, like Henry David Thoreau, wish to know an "entire earth," and if we agree that this means an Earth prior to human intrusion, then in North America, for example, we must reckon with mammoths, mastodons, camels, cheetahs, lions, ground sloths, and other lost megafauna. These are the evolutionary legacy of America. They are what is natural. If we do not consider them, we sell the continent short. To ignore the fossil record of near time in approaching conservation is as bad as a historian consulting only those books and documents written since 1 AD, ignoring earlier material, Herodotus and the Old Testament included, as too old to be of value today.

Perhaps some people feel the hemisphere is a better place for us now that the great beasts are gone. I view the loss as horrendous. It deprives us of a full measure of the wildness along with the evolutionary potential of the Americas. Without the large mammals, the land is tame; much of the emotion of the out of doors is drained.

The human species has a profound interest in and attachment to large animals. As visitors to zoos or natural history museums (where the extinct mammals occupy prime display space), circuses, rodeos, racetracks, national parks, or game parks; as hunters or animal rights activists; as watchers of movies featuring animals, or of nature documentaries on TV; as cowboys or equestrians; as dairy farmers or livestock breeders, we are, even in this high-tech age, drawn to these animals—and the bigger the better. Zoos are eager to have as many large animals on display as possible; these are their drawing cards. And the merest glimpse of a moose or bear is the high point of a wilderness excursion.

This fascination should come as no surprise. After all, for several million years our ancestors evolved in lands rich in megafauna. We were both hunters and the hunted. We followed animal trails and learned from them what plants we could eat and where fresh water could be found. Whatever our present ethnic or racial identity may be, for over a million years we shared the Quaternary stage with large animals. The large animals sported fur, hair, or a thick hide; they were warm-blooded and bright-eyed, and they nursed their young. Some traveled in herds. All were playful when young, alert when alarmed, and devoted when parents. They possessed sharp eyesight, or a keen sense of smell, or acute hearing, or all three. In these and many other ways, they would have been very much like us. We too are large mammals. It is no wonder that we are attracted to them even when, as in the case of the large carnivores, we fear them.

Toward the end of a long walk, alone at dusk, on a faint path through dark and unfamiliar woods, far from domestic sounds of town or farmhouse, a large and unexpected object looming up in the twilight, furred by what finally are seen to be mosses and lichens, brings one up short, pulse racing, inner voices chattering:

"Wait! What's that? Could it . . . is it alive?"

"Well no, of course not! Look! It's just a boulder. You're imagining things!"

And we are relieved—but also ever so slightly disappointed.

Besides our keen interest in experiencing "nature," especially large animals, a likely legacy, some claim, of our sociobiology, there is a variety of other reasons to consider near-time restorations. Some of these are broadly philosophical. For instance, it could be argued that taxa have an inherent moral right to continue evolving free of human intervention, or even that the Earth as a whole has a right to demonstrate its fullest possible evolutionary potential. It could be argued that, as the species responsible for the extinction of so many taxa, humans have a corresponding responsibility to attempt their restoration when feasible. Like all sweeping philosophical and ethical arguments, these are open to intense debate.

On the scientific front, it could be argued that existing ecosystems would be healthier and more balanced if they included their "original" complement of animal species. This argument is closely aligned with the "natural is better" philosophy discussed earlier and is essentially the position that underlies much land restoration activity today (although, of course, I would define "original" using a different time scale than would most wildlands managers). Again, when it comes to animals long extinct, counterarguments arise; for example, if ecosystems can be considered "natural" without dinosaurs, why do they need mammoths? The answer is that we are no longer in the Mesozoic; large mammals evolved in the Cenozoic. We, the dominant species of large mammal, now control the destiny of all the rest. A final argument in favor of near-time restoration efforts is that seeing the ecosystemic effects of near-time fauna would be a valuable addition to our knowledge of community ecology.

All of these arguments for resurrection have far more strong and weak points than I can examine here, and in some cases they may conflict with each other. For example, a resurrection scenario carefully tailored for scientific research might be quite different from one designed to maximize the capability of a lineage to continue its "natural" evolution. Efforts at near-time restoration would also raise a variety of important practical issues. How would restored or reintroduced species affect existing animal species, plant communities, and humans? How should we balance the interests of all these ecosystem participants?

Like the GCNP managers with their burros, some may now view as aliens the animals that evolved in America only to disappear from our land when humans arrived. They may contend that the negative impacts of recently introduced animals such as Japanese beetles, English Sparrows, and zebra mussels argue against bringing in taxa not historically present. However, relatives of large mammals once present here, such as horses, might not have such negative impacts, because other native animals and plants coevolved with them for millions of years. For the same reason, proxies for those animals might not cause problems. If they did result in ecological shifts, it would be important to examine whether these brought the affected ecosystems closer to a near-time state, if farther from the current or the immediately pre-Columbian state, and once again to consider which of these we define as "natural" or otherwise most desirable.

As is probably clear by now, I start from the general viewpoint that nature has value, that it is worthy of human preservation efforts, and that it makes the most sense to focus these efforts on what nature has looked like on our watch—during the evolutionary history of modern humans. Even with all the complexities and potential difficulties of resurrection, I believe it is worth broadening our definition of the "natural" and opening our minds to the bold new vision of conservation and wildlife management that this broader view allows. I present below several possible options for resurrection, some limited in both scope and potential consequences, and others more venturesome on both fronts.

How can it even be possible to resurrect the lost animals? Surely, if the dead cannot be brought back to life, neither can the extinct. To some extent, of course, this is true: even with radical efforts at restoration, nature can never be the same again. The key to the puzzle, however, lies in the survivors. In some cases fairly close taxonomic relatives of the lost fauna still exist, sometimes in distant corners of the planet. In others, we can at least turn to guild members, animals that are unrelated but occupy comparable trophic niches (see Flannery 2001). These survivors, some of which are themselves endangered in their current locales, may be sources of restoration, resurrection, or some approximation thereof.

Where there are closely related survivors, the approach is simple (if, on occasion, politically problematic). Especially when fossil records indicate extinction within the last few thousand years, any surviving populations of taxa on the brink of extinction should be spread back into their former ranges by any means possible. Even relatively modest efforts in this direction may have gratifying results. For example, the giant tortoises of Aldabra could be restocked in Madagascar, the Comoros, and other small islands in the Indian Ocean, as has been done in at least a few cases. Galapagos tortoises could be returned to islands emptied of giant tortoises, as has been done on Santa Fe, where the original population was eliminated by whalers and pirates. A more venturesome reintroduction would be to return giant tortoises from the cactus-clad islands of the Galapagos to similar habitat in Ecuador. This is an opportunity for restoration of one taxon of continental megafauna, at least in protected reserves.

Some may object that the introduced tortoises are not genetically identical to those eliminated hundreds of years ago. It is, of course, possible that the reintroduced animals will not prove adaptable. However, I would argue that these reintroductions are the closest we can come to true resurrection, and infinitely better than continual attrition.

Our best opportunities to save remnants of Oceania's depleted avifauna may lie in uninhabited, mainly forested islands or islets with few or no introduced species. Steadman (in Steadman and Martin 2003) recommends four translocations that he believes would stand a decent chance of success. First, he would move the endangered Marquesas Lorikeet *(Vini ultramarina)* from Ua Huka (the only island with a large population of this lorikeet) to another favorable island in the Marquesas, Fatu Hiva. Prehistoric bones show that *V. ultramarina* was widespread in the Marquesas at human arrival. The Ua Huka population itself is based on birds brought from Ua Pou in 1941.

His second suggestion involves the Polynesian megapode *(Megapodius pritchardii)*, which the fossil record indicates was widespread in Tonga at human arrival. (Megapodes, also known as bush turkeys or mound builders, once occurred much more widely in the central Pacific than they do now; Steadman 1995, n.d.) The chicks and eggs of this species were moved in 1992–1993 from its last stand on the inhabited volcanic island of Niuafo'ou to the well-forested, uninhabited volcanic islands of Late and Fonualei, where the bird still survived in 1999. The Kingdom of Tonga established megapodes on uninhabited islands by placing fertile fresh megapode eggs in mounds of rotting vegetation, their natural incubation grounds. Steadman would also release the Tooth-billed Pigeon *(Didunculus strigirostris)* on the uninhabited, steep, forested volcanic island of Tofua (which is 47 square kilometers in area), also in Tonga, even though Tofua has no fossil record of prehistoric birds. This pigeon survives only in Samoa, where it is threatened by massive deforestation.

Steadman also suggests releasing the Guam Rail *(Gallirallus owstoni)* on Aguiguan, an uninhabited, cliffy, mostly forested limestone island (7 kilometers square with an elevation of 157 meters, located 150 kilometers north-northeast of Guam). At Ionia Rockshelter, a cultural site on Aguiguan (with radiocarbon dates to 1,870 years), the most common species of bird recovered was a similar flightless rail *(Gallirallus* undescribed ssp.). The Guam Rail was lost in the wild in the 1980s but thrives in captivity. Attempts to introduce it on nearby Rota have failed thus far. Aguiguan differs from Rota in lacking people, cats, dogs, black rats, a wharf, and other major threats to the survival of rails. Given the thousands of populations of flightless rails lost in the Pacific, the success of the U.S. Fish and Wildlife Service in propagating Guam Rails may provide opportunities for undreamt-of recovery of an otherwise ill-fated group of tiny birds. The lost populations may have included many closely related taxa, some conspecific, others on the verge of or having attained their own specific status. Though distributing the Guam Rail cannot restore these lost taxa, it would be a big step toward restarting the rail evolution that has been so terribly blighted since prehistoric times.

Even on large islands that are fairly heavily settled or cultivated, the fossil record may reveal new avenues to biotic restoration. From his sampling of Holocene deposits on Kauai, the westernmost large island in Hawaii, David Burney and his colleagues have uncovered information that improves opportunities for reintroduction or restoration of species. Unconcerned with the fossil record, managers have followed biological uniformitarianism, believing that "the present is the key to the past." But when the past is not ancient and change has been anthropogenic, often it is the past that is the key to the present. In Hawaii, Burney has found prehistoric fossils less than 2,000 years old in habitats and at elevations not known to be occupied by the same species today (Burney and others 2001). Based on this record, Laysan Ducks, for example, need not be confined to the Laysan Islands, where they are vulnerable to extinction by tsunamis. They can be introduced more widely at elevations well above the reach of the most threatening tsunamis. *Pritchardia* palms can be planted, and will recover a natural range, well above the elevations that they are known to have occupied historically. Ascertaining the natural (prehuman) range of both plants and animals may help us assess where these species might thrive today.

During the cold war the Fish and Wildlife Service took the first step in megafaunal restoration not in America but in Asia. They delivered Alaskan musk oxen to Siberia to reestablish breeding herds there. Musk oxen lived in Asia until their comparatively late extinction on the Arctic coast of Siberia around 3,000 years ago. Recently Sergei Zimov has started a Quaternary Park in Siberia, planning to add woodland bison from Athabaska, Canada, to Siberian ponies and musk oxen. Zimov expects that under heavy use unpalatable plants such as Sphagnum mosses and heaths (Ericaceae), of little forage value for large mammals, will be torn up, trampled, and manured, to be replaced by more productive steppe tundra of subarctic grasses, a community that vanished locally with the extinction of mammoths (Stone 2001; Zimov et al. 1995; Zimov 2005). Some vertebrate paleontologists have proposed that climatic change forced the shift in palatability. Zimov's experiment tests the opposite possibility, that extinction of woolly mammoths, subarctic horses, and other megaherbivores triggered the shift to less palatable plant communities. His thesis appears to be testable, and if so, it certainly merits testing.

Closer to home, we have already discussed some actual and potential resurrections in North America. In its support for one such effort, the Wild Free-Roaming Horses and Burros Act, the public had it right. The elimination of wild equids around 13,000 years ago should not close the land to their surrogates. Wild equid preserves in the United States in fact represent an unintended restarting of equid evolution on the continent of equid origin. Other near-time restorations in North America have included bison, condors, and African artiodactyls. In addition, javelina, armadillo, and opossum have been moving back naturally into regions formerly inhabited by their ancestors or close relatives. South American llamas are again gaining a foothold in the pastures and rangelands of the United States and Canada. Flamingoes in the Americas, their distribution blighted by prehistoric human invaders raiding nesting colonies on the margins of salt lakes, are also a candidate for restoration. Rediscovery of an unknown wild America is under way.

Many other possible North American restorations would further broaden both the variety of animals on the continent and the ranges those animals, or their relatives, currently inhabit. I believe it is time to take an approach that includes not only creatures traditionally considered "at home on the range" but also some of those not seen roaming the Americas by any humans since the Clovis people. Some of these creatures no longer have close living relatives; in these cases we should consider proxies for them that perform the same ecological functions. The Bering Land Bridge should not be shut down forever in the interest of imagined faunal purity.

What we need are Quaternary parks, places where we seek to restore some part of the state of nature predating human impact. This could mean establishing near-time proxies in small, fenced, closely monitored research areas; releasing them in large new national parks or on private game ranches; or doing some combination of these things, or anything in between. In North America hundreds of thousands of square miles are mainly occupied by free-ranging domestic livestock, which are becoming less popular. These lands are suitable for restarting the evolution of other megaherbivores. Any specific proposal will raise its own set of ecological and political issues. Here I will sketch some general ideas in hopes of sparking in-depth discussion of the possibilities.

What are some of the animals we might consider in seeking to create a more truly natural assemblage of American fauna? Prior to 13,000 years ago the most common fossils include those of equids, camelids, and bovids. Accordingly, these are among the most important groups to consider in attempting restoration. A highly suitable bovid, the bison, is already here, with a splendid accompanying nonnative bovid, the gemsbok or fringe-eared oryx. Like the equids and camelids for which they are proxies, oryx and other grazers and browsers can teach us much about "natural" rangelands. The equids themselves are represented by wild horses and burros. Other equid species should also be considered. Equids coevolved with plants and other animals in North America for 50 million years; a large reserve for all remaining equid species would begin to remedy their 10,000-year absence. Domestic camels and llamas would be fairly close surrogates for the extinct camelids.

Other possibilities might include both the browsing black and the grass-eating white rhinoceros. The family Rhinocerotidae did not live here in near time but did so until seven million years ago. Perhaps grass-eating rhinos were replaced ecologically in the New World by grass-eating mylodons. If so, they could serve as functional surrogates for the absent ground sloths (as well as for the glyptodonts, which were probably also grazers).

Restoration of our large native carnivores would be more controversial, though perhaps not quite as inconceivable as it might at first appear. Farsighted conservationists have begun to propose "rewilding" as the foundation of a continental conservation strategy. Central to this proposition is the ongoing recovery of top predators such as grizzlies, cougars, and wolves in large parts of their native range. Although the ecological importance of such predators is now widely recognized, there has been controversy over their reintroduction in areas also used by humans, livestock, and pets. Yet an assemblage of herbivores alone cannot approximate anything like a natural balance. Beyond the species already present, obvious candidates for carnivore reintroduction would be the African lion and cheetah, close relatives of the extinct American lion and cheetah. A decade ago, conservation biologist Michael Soulé noted, "I would not be surprised to read someday that cheetahs are helping to control deer" (Soulé 1990; Owen-Smith 1989). In reality, however, bringing these animals to the open range to hunt antelope or deer, however natural it would be in evolutionary terms, seems highly unlikely.

The ultimate in American rewilding would be restoring relatives of the most influential of the missing species, those likely to have exerted the greatest influence on their natural environment. Based on what is known of living megaherbivores in Africa and Asia, and on the fossil record of the New World, there is one clear choice, animals as potent as fire in their dynamic influence on ecosystems. If we want the "super-keystone species" (Shoshani and Tassy 1996), second only to our own in its capability for altering habitats and faunas (Buss 1990; Sukumar 1994), we should include the living proboscideans—the African and Asian elephants (Martin and Burney 1999).

Based on numbers of fossils, the most common large animals in the Quaternary may have been Columbian mammoths and their close relatives, imperial mammoths. (It was the plants in the mammoth dung in Bechan Cave, plants identified by Owen Davis, that started me thinking about resurrection ecology.) American mastodons were also present in less arid parts of the West. All are gone now, of course, along with the gomphotheres in the tropics. But living African *Loxodonta africana* and Asian *Elephas maximus* are in the same order, the Proboscidea, and some taxonomists have considered *Elephas* and *Mammuthus* to be closely related (congeneric) species. An Asian elephant living today in Thailand is more akin to the extinct mammoths of North America than to living African elephants.

Introduction of both Asian and African elephants could therefore restart the evolution of proboscideans in the New World, where a dozen species held sway for more than 15 million years before their demise ended long-established ecological relationships and evolutionary possibilities. The claim of these giants to an evolutionary future is no less valid than our own. We can enable them to reinvent their ecology on the continent that once constituted an important part of their global range. For a New World elephant park suitable for wide-ranging family units, I suggest anthropogenic savannas in Central or South America, rangeland now devoted to livestock.

This restoration could also help save the African and Asian elephants. Thanks to a surging human population and to poaching for ivory, elephant numbers have crashed in the last century, and they are now at risk in many parts of their historic range. African elephants are estimated to number 550,000 to 650,000 (Douglas-Hamilton and Michelmore 1996), wild Asian *Elephas* only 37,500 to 54,600 (Sukumar and Santiapillai 1996). Saving these elephants can mean more than helping them in lands where they were known historically. Second-growth tropical rain forest in the New World might serve as reserves for them, helping the species survive as well as letting them substitute in seed dispersal for extinct South American proboscideans, ground sloths, and other lost creatures.

While I doubt she was thinking of the New World, the words of researcher Cynthia Moss are compelling: "I have realized that more than anything else, more than scientific discoveries or acceptance, what I care about and what I will fight for is the conservation, for as long as possible, not of just a certain number of elephants, but of the whole way of life of elephants. My priority, my love, my life are the Amboseli elephants, but I also want to ensure that there are elephants in other places that are able to exist in all the complexity and joy that elephants are capable of" (Moss 1988). Surely the Americas deserve to be considered among these "other places" Moss envisions. With some ranches already occupied by surplus or overage circus elephants, the New World tropics deserve to become lands of opportunity for African and Asian animals.

There, Asian *Elephas*, African *Loxodonta*, or both could show us some of the coevolutionary secrets of fruit dispersal. Beyond Quaternary parks we need Quaternary proving grounds, places to fathom as well as to celebrate our lost wildness. Dan Janzen has urged allowing Asian or African elephants to ingest, transport, and disperse, if they will, the large fruits of the American tropics (Barlow 2000); from this we could learn about the prehuman nature of tropical ecosystems, including how fruits were dispersed. Elephants also deserve a chance to taste North American vegetation. In Africa they are known to favor Bermuda grass and tamarisk, both alien to the Americas but now common along the Colorado River in Arizona and the Rio Grande drainage in New Mexico and parts of Texas.

The establishment of free-ranging elephant herds in the New World would give us unusual opportunities to learn about how nature works. What, for example, are the relationships among elephants, vegetation, and wildfire? Long smitten with the concept of a forest primeval (the climatic climax of Clementsian ecologists), North American conservation biologists have now adopted a more flexible concept of multiple stable states or discordant harmonies (Botkin 1990; Drury 1998). Conservation ecologist Graeme Caughley (1976) found no attainable natural equilibrium between elephants and forests in eastern and southern Africa. More recently Sinclair (1995) reported that African elephants and fire reach multiple stable states. Introduced elephants could teach us a great deal about the dynamic nature of wild America in evolutionary time. In their absence, inferences about the dynamics of American vegetation types could be as one-sided as those made in the absence of fire. Clearly ecologists are in danger of suffering blind spots if the largest and most potent megaherbivores long native to the Americas are missing.

We also have much to learn about the interrelationships of various animal species before historic time, and the resulting effects on vegetation. For example, on prairies between woodlands at Wind Cave National Park in South Dakota, ecologists study the interrelationships among short grasses, fire, and grazing by a free-ranging herd of bison. But bison are a small (and geologically recent) part of the pre-extinction Wild West, as we have seen. What might we learn if elephant family units were mixed in with them?

When African elephants dig for water in the dry season, the water holes they leave behind attract other species. The elephants also thin out dense stands of low trees and shrubs, in the process improving forage for other grazers (Owen-Smith 1988; Buss 1990). According to David Western, "In [Kenya's Amboseli Park] . . . you see herds of cattle filing into the park to graze, passing elephants headed out to browse. With elephants and cattle transforming the habitat in ways inimical to their own survival but beneficial to each other, they create an unstable interplay, advancing and retreating around each other like phantom dancers in a languid ecological minuet playing continuously over decades and centuries. Habitats oscillate in space like a humming top, driving and being driven by climate, animals, and people" (Western 1997, 229). In the New World we can see if bison and elephants too will dance this minuet, to the benefit of the American range.

The idea of elephants on American rangelands will likely shock and confound many conservationists and naturalists, not to mention ranchers and other devotees of the wide-open spaces. Because elephants have not lived here for over 10,000 years, many may be concerned about the ecological and practical effects of introducing these "aliens." As we have already discussed, however, the ecological impacts of creatures that evolved here are likely to be quite different than those of true aliens. A broader perspective on what is "natural," together with a focus on the opportunity to gain new insight into coevolution of vegetation and megafauna, could therefore transcend this objection. This is not the same as introducing goats or pigs onto an oceanic island whose native plants lost long ago—millions of years ago, perhaps—whatever defenses they once had to protect themselves against the tongues and teeth of large herbivores. The Americas harbored many kinds of elephants for millions of years until the last quarter of near time.

Apparently misreading my 1999 paper, some scholars have taken the trouble to warn colleagues that my views on the cause or causes of near-time extinction underlie my positive attitude toward experimental game parks in the New World. But we need not agree on what caused the extinctions to begin to take steps to remedy the damage; we need only be aware of and value what we have lost.

In the long pull all species are doomed to extinction, just as death is the inevitable fate of all individuals. Most species that once lived on Earth are no more. But this is a poor excuse for turning our back on the extraordinary loss of flagship species on our watch. Whatever happened to sweep away this rich fauna, we should work to sweep some of it back. The chance of a recurrence of whatever may have caused the extinctions need not and should not impede us in this effort. At stake are the complexity, joy, and whole way of life not only of elephants but also of a number of other surrogate species of large animals, the advance guard in the ultimate restoration of our wild lands.

EPILOGUE

Even the most ambitious rewilding efforts will not bear full fruit for many years to come. But there are other ways to acknowledge both our losses and the achievements of our species.

For starters, it is time to mourn our dead. The extinction of a few dozen species in North America in near time is trivial in a global fauna of unknown millions. What hurts is the quality of the losses and their late occurrence. The emotional impact of extinction of large warm-blooded mammals, species so late in the fossil record and so much like us, is extreme. The same can be said for flightless rails and megapodes or native pigeons *(Ducula)* on oceanic islands. Our world is far poorer for the lack of these creatures, and their loss is profoundly important, not only a tragedy to mourn but an event to commemorate. We need a Megafauna Extinction Day. In 1999, for example, the townspeople at Hot Springs, South Dakota, together with the board and staff of the Mammoth Site, held a memorial service for the extinction of mammoths. The mourners included the honey locust, *Gleditsia triacanthos,* whose pods no longer can attract native species of American elephants. Mesquite bean pods serve now as seasonal food for cattle and horses, in the absence of native camelids, horses, and mammoths.

In addition to days of remembrance, we need places of remembrance. At least some of the sites that have taught us so much about near time should also be treated as sacred places at which we honor the extinct. The Mammoth Site is an especially appropriate place for such treatment, as it was for a wake. In this paleoecological cathedral, over 100,000 visitors a year pay a modest admission fee to marvel at a unique *in situ* exhibit of splendidly preserved mammoth bones being excavated from the most concentrated natural deposit of mammoths ever exhibited on the continent. Larry Agenbroad and Jim I. Mead of Northern Arizona University and their teams of Earthwatch and other skilled volunteers have uncovered some fifty individual mammoths of two species, plus bones of the giant bear *Arctodus,* all dating from roughly 26,000 years ago. Remarkably well preserved crania are displayed in place, an extraordinary feature rarely seen even by vertebrate paleontologists. There is even a bladder stone (possibly a kidney stone) of a mammoth. Most of the mammoths are subadult males that had presumably left their herd. Experienced matriarchs leading their offspring would have seen danger in the slippery slopes above the warm waters and kept their progeny at a safe distance from the sinkhole. The dimensions of the unexcavated portion of the sinkhole suggest that another fifty young male mammoths remain to be discovered, along with other Quaternary fossils.

To my knowledge never before have professional paleontologists and townspeople cooperated as productively as they have in this case. We need more such self-sustaining public exhibits of Quaternary fossils. The attractive building protecting the outcrop is a monument to community-based education and to public fascination with America's extinct ice age mammals. The Mammoth Site adjoins herds of bison in Wind Cave National Park, Custer State Park, and the Lakota Sioux Reservation, with its fossil-rich badlands.

Many other fossil locations would be suitable sites for remembrance, such as Rancho La Brea and its magnificent Page Museum in Hancock Park, Los Angeles. Caves, too, deserve respectful treatment for their fossil deposits, especially those few that contain manure of extinct ground sloths and other Quaternary animals, thus enabling us to study both the ecology and the natural history of a remarkable group of extinct animals. Both Stanton's Cave and Rampart Cave, for example, are sepulchers not simply for the dead but for the extinct. Extinct ground sloth remains are incorporated in the interpretation of Kartchner Caverns, Arizona, and Grand Canyon Caverns on Route 66 in northwestern Arizona.

At one or more of these or some other locations, a memorial would be appropriate. I think of an arch like that at St. Louis, or a cenotaph like the Washington Monument, or perhaps a shining black wall inscribed with the names of all the species lost in near time up to the present. In any case, the center I envision needs a fundamentally new design appropriate to the subject.

It would also be appropriate to pay more attention to how our museums treat extinct megafauna. Beyond the ever-popular dinosaur exhibits, one usually finds a gallery of excellent reconstructions of the skeletons of some of the large extinct late-Quaternary mammals. The Page Museum and the Los Angeles County Museum have particularly good collections and outstanding exhibits. Visiting them is a must in any effort to appreciate what suffered extinction in western North America. But many of our great museums display reconstructed animals in dioramas of presently game-rich parts of the planet, such as Africa, while America's preextinction megafauna appear only as skeletons, without natural backgrounds of their American habitats.

Cost, of course, is a factor; a simulated woolly mammoth, life size, with an artificial coat of hair from other animals, may cost $10,000. But perspective plays a role as well. For example, Tucson's International Wildlife Museum of the Safari Club displays a woolly mammoth against a background of treeless tundra. When the display was being prepared, I complained to the curator that woolly mammoths did not range into Arizona and suggested an exhibit of ice age Arizona, hopefully one that would include my favorite, the Shasta ground sloth. "Where can they be obtained?" was the reply. I had no idea.

Someday, however, I hope that the six-year-olds who can rattle off fifteen generic names of dinosaurs will start mastering such strange names as *Castoroides* for the giant beaver in the Midwest, *Cuvieronius* for an extinct mastodon in tropical America, *Eremotherium* for an extinct giant ground sloth in Florida, and *Platygonus* for an extinct giant peccary in Missouri. These and other extinct large mammals and birds have much more to teach us about our own remarkable past, and about designing for the future, than do the dinosaurs.

Our own ancestors, too, should be commemorated. Native Americans and others alike need to honor the achievement of the First Americans in colonizing the largest uninhabited landmass ever discovered by *Homo sapiens*. There is growing international interest in this extraordinary feat, but it is largely neglected in our own vision of our history. A hemispheric holy day, a Clovis Day, need not replace Columbus Day in commemorating the discovery of America; it would simply place that discovery in perspective.

We also need monuments, as well as museums and interpretive cen-

ters, at some, if not all, of the important early North American archaeological sites, especially those that also include extinct mammals. Beyond
the Page Museum at Rancho La Brea, good candidates for such memo-
rials include the Clovis-mammoth sites along the San Pedro River in
Cochise County, Arizona; Blackwater Draw, near Clovis, New Mexico;
the Colby site in Wyoming, where bones suggesting a frozen meat cache
were found on top of a Clovis point; and the East Wenatchee site at which
a cache of ceremonial (oversized) Clovis artifacts was found buried in
an apple orchard in Washington. From the perspective of both America's
first pioneers and the extinct species, all these are sacred grounds.

If controversial claims such as Meadowcroft, Monte Verde, or Bluefish
Caves can be replicated, of course pre-Clovis invasion of the Americas
will be equally worth celebrating. The arrival of pre-Clovis humans might
be viewed as the Paleolithic equivalent of the Vikings' reaching America
in advance of Columbus.

In the end, though, commemorations and monuments only touch the
surface of the rebirth that is needed. I can think of no better or more important monument to the discovery of America than efforts at restarting
the evolution of our extinct fauna. It is already under way wherever free-
ranging horses or wild burros roam the range.

References

Abdullahi, A. N. 1980. Forage preference of cattle and eland on a desert grassland in southern Arizona. Ph.D. diss., University of Arizona.

Adavasio, J. M., and D. R. Pedler. 1997. Monte Verde and the antiquity of humankind in the Americas. *Antiquity* 71:573–580.

Adavasio, J. M., with J. Page. 2002. *The first Americans: In pursuit of archaeology's greatest mystery.* New York: Random House.

Agenbroad, L. D., and J. I. Mead, eds. 1994. *The Hot Springs mammoth site: A decade of field and laboratory research in paleontology, geology, and paleoecology.* Hot Springs: Mammoth Site of South Dakota.

———. 1996. Distribution and palaeoecology of central and western American *Mammuthus.* In J. Shoshani and P. Tassy, eds., *The Proboscidea: Evolution and palaeoecology of elephants and their relatives,* 280–288. Oxford: Oxford University Press.

Alcock, J. 1993. *The masked bobwhite rides again.* Tucson: University of Arizona Press.

Alley, R. B. 2000. *The two-mile time machine: Ice cores, abrupt climatic change and our future.* Princeton: Princeton University Press.

Alroy, J. 1999. Putting North America's end-Pleistocene megafaunal extinction in context: Large-scale analyses of spatial patterns, extinction rates, and size distributions. In R. D. E. MacPhee, ed., *Extinctions in near time: Causes, contexts, and consequences,* 105–143. New York: Kluwer Academic / Plenum Publishers.

———. 2001. A multispecies overkill simulation of the end-Pleistocene megafaunal mass extinction. *Science* 292:1893–1896.

Alvarez, L. W., W. Alvarez, F. Asaro, and H. V. Michel. 1980. Extraterrestrial cause for the Cretaceous-Tertiary extinction. *Science* 208:1095–1108.

Alvarez, W. 1997. T. rex *and the crater of doom*. Princeton: Princeton University Press.

Anderson, A. 1997. Prehistoric Polynesian impact on the New Zealand environment: Te Whenua Hou. In P. V. Kirch and T. L. Hunt, eds., *Historical ecology in the Pacific islands: Prehistoric environmental and landscape change*, 271–283. New Haven: Yale University Press.

Anderson, A. J., and M. S. McGlone. 1992. Prehistoric land and people in New Zealand. In J. Dodson, ed., *The naïve lands: Human-environmental interactions in Australia and Oceania*, 199–241. Sydney: Longman Cheshire.

Anderson, E. 1984. Who's who in the Pleistocene: A mammalian bestiary. In P. S. Martin and R. G. Klein, eds., *Quaternary extinctions: A prehistoric revolution*, 40–89. Tucson: University of Arizona Press.

Anderson, R. S., J. L. Betancourt, J. I. Mead, R. H. Hevly, and D. P. Adam. 2000. Middle and late-Wisconsin paleobotanic and paleoclimatic records from the southern Colorado Plateau, USA. *Paleogeography, Paleoclimatology, Paleoecology* 155:31–57.

Azzaroli, A. 1992. Ascent and decline of monodactyl equids: A case for prehistoric overkill. *Annales Zoologici Fennici* 28:151–163.

Bahn, P. G., and J. Vertut. 1997. *Journey through the ice age*. Berkeley: University of California Press.

Bakeless, J. 1961 (1950). *The eyes of discovery*. New York: Dover Publications.

Baldwin, G. C. 1946. Notes on Rampart Cave. *Masterkey* 20:94–96.

Barlow, C. 2000. *The ghosts of evolution*. New York: Basic Books.

Barnosky, A. D. 1986. Big game extinction caused by late Pleistocene climatic change: Irish elk *(Megaloceros giganteus)* in Ireland. *Quaternary Research* 25:128–135.

Behrens, C. 1971. The search for New World man. *Science News* 99:98–100.

Belsky, A. J. 1992. Letter to *Fremontia* 20:30–31.

Berger, J., J. E. Swenson, and I. L. Persson. 2001. Recolonizing carnivores and naïve prey: Conservation lessons from Quaternary extinctions. *Science* 291: 1036–1039.

Betancourt, J. L., T. R. Van Devender, and P. S. Martin, eds. 1990. *Packrat middens: The last 40,000 years of biotic change*. Tucson: University of Arizona Press.

Bock, C. E., and J. H. Bock. 2000. *The view from Bald Hill*. Berkeley: University of California Press.

Boldurian, A. T., and J. L. Cotter. 1999. *Clovis revisited: New perspectives on Paleoindian adaptations from Blackwater Draw, New Mexico*. Philadelphia: University Museum, University of Pennsylvania.

Bonnichsen, R., and K. L. Turnmire. 1999. An introduction to the peopling of the Americas. In R. Bonnichsen and K. L. Turnmire, eds., *Ice age people of North America: Environments, origins, and adaptations*, 1–26. Corvallis: Oregon State University Press.

Bonnicksen, T. M., M. K. Anderson, H. T. Lewis, C. E. Kay, and R. Knudsen. 1999. Native American influences on the development of forest ecosystems. In R. C. Szaro, N. C. Johnson, W. T. Sexton, and A. J. Malk, eds., *Ecological stewardship: A common reference for ecosystem management*, vol. 2, 439–470. Oxford: Elsevier Science.

Botkin, D. B. 1990. *Discordant harmonies: A new ecology for the twenty-first century.* New York: Oxford University Press.

Boulter, M. C. 2002. *Extinction: Evolution and the end of man.* New York: Columbia University Press.

Bover, O., and J. A. Alcover. 2003. Understanding late Quaternary extinctions: The case of *Myotragus balearicus* (Bate, 1909). *Journal of Biogeography* 30:771–781.

Bowers, J. E. 1988. *A sense of place: The life and work of Forrest Shreve.* Tucson: University of Arizona Press.

Bowers, J. E., and R. M. Turner. 1985. A revised vascular flora of Tumamoc Hill, Tucson, Arizona. *Madroño* 32:225–252.

Boyd, R. 1999. *The coming of the spirit of pestilence: Introduced infectious diseases and population decline among Northwest Coast Indians, 1774–1874.* Vancouver: UBC Press; Seattle: University of Washington Press.

Brook, B. W., and M. J. S. Bowman. 2002. Explaining the Quaternary megafaunal extinctions: Models, chronologies, and assumptions. *Proceedings of the National Academy of Sciences* 99:14624–14627.

Broughton, J. 1997. Widening diet breadth, declining foraging efficiency and prehistoric harvest pressures: Ichthyofaunal evidence from the Emeryville Shellmound, California. *Antiquity* 71:845–862.

Brown, D. E., F. Reichenbacher, and S. E. Franson. 1998. *A classification of North American biotic communities.* Salt Lake City: University of Utah Press.

Brown, J. H., and M. V. Lomolino. 1998. *Biogeography.* 2d ed. Sunderland, MA: Sinauer Associates.

Budyko, M. I. 1967. On the causes of the extinction of some animals at the end of the Quaternary. *Soviet Geography—Review and Translations* 8:783–793.

Bunce, M., T. M. Worthy, T. Ford, W. Hoppitt, E. Willerslev, A. Drummond, and A. Cooper. 2003. Extreme reversed sexual size dimorphism in the extinct New Zealand moa *Dinornis. Nature* 425:172–175.

Burkhardt, J. W. 1996. Herbivory in the intermontane west: An overview of evolutionary history, historic cultural impacts and lessons from the west. College of Forestry Wildlife and Range Sciences, University of Idaho, Station Bulletin no. 58. Moscow: Idaho Forest, Wildlife and Range Experiment Station.

Burney, D. A. 1993. Recent animal extinctions: Recipes for disaster. *American Scientist* 81:530–541.

Burney, D. A., L. P. Burney, L. R. Godfrey, W. L. Jungers, S. M. Goodman, H. T. Wright, and A. J. T. Jull. 2004. A chronology for late prehistoric Madagascar. *Journal of Human Evolution* 47:25–63.

Burney, D. A., H. F. James, L. P. Burney, S. L. Olson, W. Kikuchi, W. L. Wagner, M. Burney, D. McCloskey, D. Kikuchi, F. V. Grady, R. Gage II, and R. Nishek. 2001. Fossil evidence for a diverse biota from Kaua'i and its transformation since human arrival. *Ecological Monographs* 71:615–641.

Burney, D. A., and Ramilisonina. 1998. The kilopilopitsofy, kidoky, and bokyboky: Accounts of strange animals from Belo-sur-mer, Madagascar, and the megafaunal "extinction window." *American Anthropologist* 100:957–966.

Burney, D. A., G. S. Robinson, and L. P. Burney. 2003. *Sporormiella* and the late

Holocene extinctions in Madagascar. *Proceedings of the National Academy of Sciences* 100:10800–10805.

Buss, I. O. 1990. *Elephant life: Fifteen years of high population density*. Ames: Iowa State University Press.

Callenbach, E. 1999. *Bring back buffalo! A sustainable future for America's Great Plains*. Berkeley: University of California Press.

Carothers, S. W. 1996. Feral burros: Old arguments and new twists. In R. H. Webb, *Grand Canyon, a century of change: Rephotography of the 1889–1890 Stanton expedition*, 84–85. Tucson: University of Arizona Press.

Carothers, S. W., M. E. Stitt, and R. R. Johnson. 1976. "Feral asses on public lands: An analysis of biotic impact, legal considerations, and management alternatives." *Transactions of the 41st North American Wildlife and Natural Resources Conference*. Washington, DC: Wildlife Management Institute.

Caughley, G. 1976. The elephant problem: An alternative hypothesis. *East African Wildlife Journal* 14:265–283.

————. 1988. The colonization of New Zealand by the Polynesians. *Journal of the Royal Society of New Zealand* 18:245–270.

Chauvet, J.-M., E. B. Deschamps, and C. Hillaire. 1996. *Dawn of art: The Chauvet cave; The oldest known paintings in the world*. New York: Henry N. Abrams.

Clark, F. E., W. A. O'Deen, and D. E. Belau. 1974. Carbon, nitrogen and ^{15}N content of fossil and modern dung from the lower Grand Canyon. *Journal of the Arizona-Nevada Academy of Science* 9:95–96.

Clark, R. C. 1977. Plant taxa in late Pleistocene artiodactyl fecal pellets, Rampart Cave, Arizona. Master's thesis, Colorado State University.

Coats, L. L. 1997. Middle to late Wisconsinan vegetation change at Little Nankoweap, Grand Canyon National Park, Arizona. Master's thesis, Northern Arizona University.

Coe, M. 1981. Body size and the extinction of the Pleistocene megafauna. *Paleoecology of Africa* 13:139–145.

————. 1982. The bigger they are . . . *Oryx* 16:225–228.

Cole, K. L. 1988. Past rates of change, species richness, and a model of vegetation inertia in the Grand Canyon, Arizona. *American Naturalist* 125:289–303.

————. 1990. Late-Quaternary vegetation gradients through the Grand Canyon. In J. L. Betancourt, T. R Van Devender, and P. S. Martin, eds., *Packrat middens: The last 40,000 years of biotic change*, 240–258. Tucson: University of Arizona Press.

Collins, P. W., N. F. R. Snyder, and S. D. Emslie. 2000. Faunal remains in California Condor nest caves. *Condor* 103:222–227.

Connin, S. L., J. Betancourt, and J. Quade. 1998. Late Quaternary C_4 plant dominance and summer rainfall in the southwestern United States from isotopic study of herbivore teeth. *Quaternary Research* 50:179–193.

Coope, G. R. 1995. Insect faunas in ice age environments: Why so little extinction? In J. H. Lawton and R. M. May, eds., *Extinction rates*, 55–74. Oxford: Oxford University Press.

Corbett, J. W. 1973. Late Pleistocene extinctions. Letter to *Science* 180:905.

Czaplewski, N. J., and C. Cartelle. 1998. Pleistocene bats from cave deposits in Bahia, Brazil. *Journal of Mammalogy* 79:487–803.

Dansgaard, W., S. J. Johnsen, H. B. Clausen, D. Dahl-Jensen, N. S. Gundestrup, C. U. Hammer, C. S. Hvidberg, J. P . Steffensen, A. E. Sveinbjornsdottir, J. Jouzel, and G. Bond. 1993. Evidence for general instability of past climate from a 250-kyr ice-core record. *Nature* 364:218–220.

d'Antoni, H. 1983. Pollen analysis of Gruta del Indio. In G. Rabassa, ed., *Quaternary of South America and Antarctic Peninsula,* 83–108. Rotterdam: A. A. Balkema.

Davis, O. K. 1987. Spores of the dung fungus *Sporormiella:* Increased abundance in historic sediments and before Pleistocene megafaunal extinction. *Quaternary Research* 28:290–294.

Davis, O. K., L. D. Agenbroad, P. S. Martin, and J. I. Mead. 1984. The Quaternary dung blanket of Bechan Cave, Utah. In H. H. Genoways and M. R. Dawson, eds., *Contributions in Quaternary vertebrate paleontology: A volume in memorial to John E. Guilday,* 267–282. Special Publication of Carnegie Museum of Natural History, no. 8. Pittsburgh: Carnegie Museum of Natural History.

Davis, O. K., J. I. Mead, P. S. Martin, and L. D. Agenbroad. 1985. Riparian plants were a major component of the diet of mammoths of southern Utah. *Current Research in the Pleistocene* 2:81–82.

Davis, O. K., T. Minckley, T. Moutoux, T. Jull, and B. Kalin. 2002. The transformation of Sonoran Desert wetlands following the historic decrease of burning. *Journal of Arid Environments* 50:393–412.

Dayton, P. K. 1974. Experimental studies of algal canopy interactions in a sea otter–dominated kelp community at Amchitka Island, Alaska. *Fishery Bulletin* 73:230–237.

Deevey, E. S. 1942. A re-examination of Thoreau's "Walden." *Quarterly Review of Biology* 17:1–11.

———. 1949. Biogeography of the Pleistocene. *Geological Society of America Bulletin* 60:1315–1416.

———. 1967. Introduction. In P. S. Martin and H. E. Wright Jr., eds., *Pleistocene extinctions: The search for a cause,* 63–72. New Haven: Yale University Press.

Del Giudice, G. D. 1998. Surplus killing of white-tailed deer by wolves in north central Minnesota. *Journal of Mammalogy* 79:227–235.

Deloria, V., Jr. 1995. *Red earth, white lies: Native Americans and the myth of scientific fact.* New York: Scribner.

Dewar, R. E. 1997. Were people responsible for the extinction of Madagascar's subfossils, and how will we ever know? In S. Goodman and B. Patterson, eds., *Natural change and human impact in Madagascar,* 364–377. Washington, DC: Smithsonian Institution Press.

Diamond, J. 1984. Historic extinctions: A Rosetta stone for understanding prehistoric extinctions. In P. S. Martin and R. G. Klein, eds., *Quaternary extinctions: A prehistoric revolution,* 824–862. Tucson: University of Arizona Press.

———. 1992. *The third chimpanzee: The evolution and future of the human animal.* New York: Harper Collins.

———. 1997. *Guns, germs, and steel: The fate of human societies.* New York: W. W. Norton.

DiPeso, C. C., J. B. Rinaldo, and G. J. Fenner. 1974. *Casas Grandes: A fallen trading center of Gran Chichimeca.* Vol. 7. Flagstaff: Northland Press.

Dobyns, H. F. 1983. *Their number became thinned.* Knoxville: University of Tennessee Press.

———. 1993. Disease transfer at contact. *Annual Review of Anthropology* 22:273–291.

Douglas-Hamilton, I., and F. Michelmore. 1996. *Loxodonta africana:* Range and distribution, past and present. In J. Shoshani and P. Tassy, eds., *The Proboscidea: Evolution and palaeoecology of elephants and their relatives,* 321–326. Oxford: Oxford University Press.

Drury, W. H. 1998. *Chance and change: Ecology for conservationists.* Berkeley: University of California Press.

Dudley, J. P. 1999. Seed dispersal of *Acacia erioloba* by African bush elephants in Hwange National Park, Zimbabwe. *African Journal of Ecology* 37: 375–385.

———. 2000. Seed dipersal by elephants in semiarid woodland habitats of Hwange National Park, Zimbabwe. *Biotropica* 32:556–561.

Eames, A. J. 1930. Report on ground sloth coprolite from Dona Ana County, New Mexico. *American Journal of Science,* 5th series, 20:353–356.

Edwards, S. W. 1992. Observations on the prehistory and ecology of grazing in California. *Fremontia* 20:3–20.

Emslie, S. D. 1987. Age and diet of fossil California Condors in Grand Canyon, Arizona. *Science* 237:768–770.

———. 1992. Early humans in North America. *Science* 256:426–427.

Emslie, S. D., R. C. Euler, and J. I. Mead. 1987. A desert culture shrine in Grand Canyon, Arizona, and the role of split-twig figurines. *National Geographic Research* 3:511–516.

Emslie, S. D., J. I. Mead, and L. Coats. 1995. Split-twig figurines in Grand Canyon, Arizona: New discoveries and interpretations. *Kiva* 61:145–173.

Euler, R. C., ed. 1984. *The archaeology, geology, and paleobiology of Stanton's Cave, Grand Canyon National Park, Arizona.* Grand Canyon Natural History Association, Monograph no. 6. Grand Canyon, AZ: Grand Canyon Natural History Association.

Faber, P. M. 1992. Editorial in *Fremontia* 20:2.

Fagan, B. M. 1996. *The Oxford companion to archaeology.* New York: Oxford University Press.

Fariña, R. A. 1996. Trophic relationships among Lujanian mammals. *Evolutionary Theory* 11:125–134.

Fariña, R. A., S. F. Vizcaíno, and M. S. Bargo. 1997. Body mass estimations in Lujanian (late Pleistocene–early Holocene of South America) mammal megafauna. *Mastozoología Neotropical* 5:87–108.

Farrand, W. R. 1961. Frozen mammoths and modern geology. *Science* 133:729–735.

Fayrer-Hosken, R. A., D. Grobler, J. J. Van Altena, H. J. Bertschinger, and J. F. Kirkpatrick. 2000. Immunocontraception of African elephants: A humane method to control elephant populations without behavioral side effects. *Nature* 407:149.

Ferguson, C. W. 1984. Dendrochronology of driftwood from Stanton's Cave. In R. C. Euler, ed., *The archaeology, geology, and paleobiology of Stanton's Cave, Grand Canyon National Park, Arizona*, 93–98. Grand Canyon Natural History Association, Monograph no. 6. Grand Canyon, AZ: Grand Canyon Natural History Association.

Ferring, C. R. 2001. The archaeology and paleoecology of the Aubrey Clovis site (41DN479) Denton County, Texas. Denton: Center for Environmental Archaeology, University of North Texas.

Fiedel, S. J. 1992. *Prehistory of the Americas*. Cambridge: Cambridge University Press.

———. 2003. Quacks in the ice: Waterfowl, Paleoindians, and the discovery of America. Expanded version of a paper presented at the 65th annual meeting of the Society for American Archaeology, April 6, 2000.

Fiedel, S. J., and G. Haynes. 2003. A premature burial: Comments on Grayson and Meltzer's requiem for overkill. *Journal of Archaeological Science* 31:121–131.

Fisher, D. C. 1996. Extinction of proboscideans in North America. In J. Shoshani and P. Tassy, eds., *The Proboscidea: Evolution and palaeoecology of elephants and their relatives*, 296–315. Oxford: Oxford University Press.

Flannery, T. F. 2001. *The endless frontier: An ecological history of North America and its peoples*. New York: Atlantic Monthly Press.

Flannery, T. F., and P. Schouten. 2001. *A gap in nature: Discovering the world's extinct animals*. New York: Atlantic Monthly Press.

Fleming, C. A. 1962. The extinction of moas and other animals during the Holocene period. *Notornis* 10:113–117.

Flenley, J. 1979. *The equatorial rain forest: A geological history*. Boston: Butterworths.

Flint, R. F. 1971. *Glacial and Quaternary ecology*. New York: John Wiley & Sons.

Flores, D. 2001. *The natural west: Environmental history in the Great Plains and Rocky Mountains*. Norman: University of Oklahoma Press.

Foreman, D. 2004. *Rewilding North America: A vision for conservation in the twenty-first century*. Washington, DC: Island Press.

Frison, G. C. 1998. Paleoindian large mammal hunters on the plains of North America. *Proceedings of the National Academy of Sciences* 95:14576–14583.

Frison, G., and B. Bradley. 1999. *The Fenn cache: Clovis weapons and tools*. Santa Fe: One Horse Land and Cattle Company.

Garcia, A. 2003. On the coexistence of man and extinct Pleistocene megafauna at Gruta del Indio (Argentina) 9000 C-14 years ago. *Radiocarbon* 45:33–39.

Garcia, A., and A. H. Lagiglia. 1999. A 30,000-year-old dung layer from Gruta del Indio, Mendoza, Argentina. *Current Research in the Quaternary* 16:116–118.

Geist, V. 1996. *Buffalo nation: History and legend of the North American bison*. Stillwater, MN: Voyageur Press.

Gillespie, R. 2002. Dating the first Australians. *Radiocarbon* 44:455–472.

———. n.d. Blitzkrieg versus climatic change in megafaunal extinctions: No contest. Unpublished ms.

Gillette, D. D., and D. B. Madsen. 1993. The Columbian mammoth, *Mammuthus columbi*, from the Wasatch Mountains of central Utah. *Journal of Paleontology* 87:669–680.

Gittleman, J. L., and M. E. Gompper. 2003. The risk of extinction: What you don't know will hurt you. *Science* 291:997–999.

Gonzalez, S., A. C. Kitchener, and A. M. Lister. 2000. Survival of the Irish elk into the Holocene. *Nature* 405:753–754.

Gould, S. J. 1989. *Wonderful life: The Burgess Shale and the nature of history.* New York: W. W. Norton.

Graham, R. W., and E. L. Lundelius. 1994. *Faunmap: A database documenting late Quaternary distributions of mammal species in the United States.* Illinois State Museum, Scientific Papers, vol. 25, nos. 1 and 2. Springfield: Illinois State Museum.

Grayson, D. K. 1991. Late Pleistocene mammalian extinctions in North America: taxonomy, chronology, and explanations. *Journal of World Prehistory* 5:193–232.

———. 2001. The archaeological record of human impacts on animal populations. *Journal of World Prehistory* 15:1–68.

Grayson, D. K., and D. J. Meltzer. 2002. Clovis hunting and large mammal extinction: A critical review of the evidence. *Journal of World Prehistory* 16:313–359.

———. 2003. A requiem for North American overkill. *Journal of Archaeological Science* 30:585–593.

Gregg, J. 1905. Commerce of the prairies, or The journal of a Santa Fe trader, during eight expeditions across the great western prairies . . . Reprinted in R. G. Thwaites, ed., *Early western travels, 1748–1846.* Cleveland: Arthur H. Clark.

Guthrie, R. D. 1990. *Frozen fauna of the mammoth steppe: The story of Blue Babe.* Chicago: University of Chicago Press.

———. 2001. Origin and causes of the mammoth steppe: A story of cloud cover, woolly mammoth tooth pits, buckles, and inside-out Beringia. *Quaternary Science Reviews* 20:549–574.

———. 2003. Rapid body size decline in Alaskan Pleistocene horses before extinction. *Nature* 476:169–171.

———. 2004. Radiocarbon evidence of mid-Holocene mammoths stranded on an Alaskan Bering Sea island. *Nature* 479:746–749.

———. 2005. *The nature of paleolithic art.* Chicago: University of Chicago Press.

Hall, D. A. 1998. Arizona's famous Clovis sites could be displayed for public; scientist outlines plan for preservation. *Mammoth Trumpet* 13(2):2–6, 20.

Hansen, R. M. 1978. Shasta ground sloth food habits, Rampart Cave, Arizona. *Paleobiology* 4:302–319.

———. 1980. Late Pleistocene plant fragments in the dungs of herbivores at Cowboy Cave. In J. D. Jennings, ed., *Cowboy Cave,* 179–189. Anthropological Papers, University of Utah, no. 104. Salt Lake City: University of Utah Press.

Hansen, R. M., and P. S. Martin. 1973. Ungulate diets in the lower Grand Canyon. *Journal of Range Management* 26:380–381.

Hanski, I., and Y. Cambefort, eds. 1991. *Dung beetle ecology.* Princeton: Princeton University Press.

Harington, C. R. 1972. Extinct animals of Rampart Cave. *Canadian Geographical Journal* 85:178–183.

———. 1984. The ungulate remains: An identification list. In R. C. Euler, ed., *The archaeology, geology, and paleobiology of Stanton's Cave, Grand Canyon National Park, Arizona,* 67–75. Grand Canyon Natural History Association, Monograph no. 6. Grand Canyon, AZ: Grand Canyon Natural History Association.

Harrington, M. R. 1933. *Gypsum Cave, Nevada: Report of the second Sessions expedition.* Southwest Museum Papers, no. 8. Los Angeles: Southwest Museum.

———. 1936. A new ground sloth den. *Masterkey* 10:225–227.

Harris, A. H. 1985. *Late Pleistocene vertebrate paleoecology of the west.* Austin: University of Texas Press.

———. 1993. Quaternary vertebrates of New Mexico. In S. G. Lucas and J. Zidek, eds., *Vertebrate paleontology in New Mexico,* 179–191. Bulletin (New Mexico Museum of Natural History and Science) 2. Albuquerque: Authority of State of New Mexico.

Hastings, J. R., and R. M. Turner. 1965. *The changing mile: An ecological study of vegetation change with time in the lower mile of an arid and semiarid region.* Tucson: University of Arizona Press.

Haury, E. W. 1953. Artifacts with mammoth remains, Naco, Arizona. I. Discovery of the Naco mammoth and associated projectile points. *American Antiquity* 19:1–14.

Haury, E. W., E. B. Sayles, and W. W. Wasley. 1959. The Lehner mammoth site, southeastern Arizona. *American Antiquity* 25:2–30.

Hauthal, R., S. Roth, and R. Lehmann-Nitsche. 1899. El mamífero misterioso de la Patagonia, *Grypotherium domesticum. Revista del Museo de la Plata* 9:409–474.

Haynes, C. V., Jr. 1967a. Carbon-14 dates and early man in the New World. In P. S. Martin and H. E. Wright Jr., eds., *Pleistocene extinctions: The search for a cause,* 267–286. New Haven: Yale University Press.

———. 1967b. Quaternary geology of the Tule Springs area, Clark County, Nevada. In H. M. Wormington and D. Ellis, eds., *Pleistocene studies in southern Nevada.* Nevada State Museum Anthropological Papers, no. 13. Carson City: Nevada State Museum.

———. 1991. Geoarchaeological and paleohydrological evidence for a Clovis-age drought in North America and its bearing on extinction. *Quaternary Research* 35:435–450.

———. 1993. Contributions of radiocarbon dating to the geochronology of the peopling of the New World. In R. E. Taylor, A. Long, and R. S. Kra, eds., *Radiocarbon dating after four decades,* 355–374. New York: Springer-Verlag.

Haynes, G. 2001. The catastrophic extinction of North American mastodonts. *World Archaeology* 33:391–416.

———. 2002. *The early settlement of North America: The Clovis era.* Cambridge: Cambridge University Press.

Heinrich, B. 1999. *Mind of the raven: Investigations and adventures with wolf birds.* New York: Cliff Street Books.

Heizer, R. F., and R. Berger. 1970. Radiocarbon age of Gypsum Cave culture. *Contributions of the University of California Archaeological Research Facility* 7:13–18.

Henry, M. 1953. *Brighty of the Grand Canyon.* New York: Rand McNally.

Hereford, R. 1984. Driftwood in Stanton's Cave: The case for temporary damming of the Colorado River at Nankoweap Creek in Marble Canyon. In R. C. Euler, ed., *The archaeology, geology, and paleobiology of Stanton's Cave, Grand Canyon National Park, Arizona,* 99–106. Grand Canyon Natural History Association, Monograph no. 6. Grand Canyon, AZ: Grand Canyon Natural History Association.

Hibbard, C. W. 1958. Summary of North American Pleistocene mammalian local faunas. *Michigan Academy of Science Papers* 43:3–32.

Hickerson, H. 1970. *The Chippewa and their neighbors: A study in ethnohistory.* New York: Holt, Rinehart, and Winston.

Higuchi, H., K. Ozaki, G. Fujita, J. Minton, M. Ueta, M. Soma, and N. Mita. 1996. Satellite tracking of white-naped crane migration and the importance of the Korean Demilitarized Zone. *Conservation Biology* 10:806–812.

Hoffmeister, D. F. 1986. *Mammals of Arizona.* Tucson: University of Arizona Press and Arizona Game and Fish Department.

Hofreiter, M., J. L. Betancourt, A. Pelliza Sbriller, V. Markgraf, and H. G. McDonald. 2003. Phylogeny, diet, and habitat of an extinct ground sloth from Cuchillo Cura, Neuquén Province, southwest Argentina. *Quaternary Research* 59:364–378.

Holdaway, R. N. 1999. A spatio-temporal model for the invasion of the New Zealand archipelago by the Pacific rat *(Rattus exulans). Journal of the Royal Society of New Zealand* 29:91–105.

Holdaway, R. N., and C. Jacomb. 2000. Rapid extinction of the moas (Aves: *Dinornithiformes)*: Model, test, and implications. *Science* 287:2250–2254.

Holdaway, R. N., R. G. Roberts, N. R. Bevan-Athfield, J. M. Olley, and T. H. Worthy. 2002. Optical dating of quartz sediments and accelerator mass spectrometry ^{14}C dating of bone gelatin and moa eggshell: A comparison of age estimates for non-archaeological deposits in New Zealand. *Journal of the Royal Society of New Zealand* 32:463–505.

Holloway, M. 1999. Beast in the mist. *Discovery* (September):58–65.

Holman, J. A. 1995. *Pleistocene amphibians and reptiles in North America.* New York: Oxford University Press.

Hoss, M., A. Dilling, A. Currant, and S. Paavo. 1996. Molecular phylogeny of the extinct ground sloth *Mylodon darwinii. Proceedings of the National Academy of Sciences* 93:181–185.

Houk, R. 1996. *An introduction to Grand Canyon ecology.* Grand Canyon, AZ: Grand Canyon Association.

Hughen, K. A., J. T. Overpeck, S. J. Lehman, M. Kashgarian, J. Southon, L. C. Peterson, R. Alley, and D. M. Sigman. 1998. Deglacial changes in ocean circulation from an extended radiocarbon calibration. *Nature* 391:65–68.

Hulbert, R. C., Jr., ed. 2001. *The fossil vertebrates of Florida.* Gainesville: University of Florida Press.

Huntley, B., W. Cramer, A. V. Morgan, H. C. Prentice, and J. R. M. Allen. 1997.

Past and future rapid environmental changes: The spatial and evolutionary responses of terrestrial biota. Berlin: Springer-Verlag.

Huynen, L., C. D. Millar, R. P. Scofield, and D. M. Lambert. 2003. Nuclear DNA sequences detect species limits in ancient moa. *Nature* 425:175–178.

Iturralde-Vinent, M. A., and R. D. E. MacPhee. 1999. Paleogeography of the Caribbean region: Implications for Cenozoic biogeography. *Bulletin of the American Museum of Natural History* 238:1–98.

Iturralde-Vinent, M. A., R. D. E. MacPhee, S. Díaz-Franco, R. Rojas-Consuegra, W. Suárez, and A. Lomba. 2000. La Brea de San Felipe, a Quaternary fossiliferous asphalt seep near Martí (Matanzas Province, Cuba). *Caribbean Journal of Science* 36:300–313.

Jackson, S. T., and C. Weng. 1999. Late Quaternary extinction of a tree species in eastern North America. *Proceedings of the National Academy of Sciences* 96:13847–13852.

Janzen, D. 1983. The Pleistocene hunters had help. *American Naturalist* 121: 598–599.

Janzen, D., and P. S. Martin. 1982. Neotropical anachronisms: The fruits the gomphotheres ate. *Science* 215:19–27.

Jelinek, A. J. 1967. Man's role in the extinction of Pleistocene faunas. In P. S. Martin and H. E. Wright Jr., eds., *Pleistocene extinctions: The search for a cause*, 193–200. New Haven: Yale University Press.

Jennings, J. D., ed. 1980. *Cowboy Cave.* Anthropological Papers, University of Utah, no. 104. Salt Lake City: University of Utah Press.

Kay, C. E. 1994. Aboriginal overkill: The role of Native Americans in structuring western ecosystems. *Human Nature* 5:359–398.

———. 1998. Are ecosystems structured from the top-down or bottom-up? A new look at an old debate. *Wildlife Society Bulletin* 26:484-498.

———. 2002. Are ecosystems structured from the top-down or bottom-up? A new look at an old debate. In C. E. Kay and R. T. Simmons, eds., *Wilderness and political ecology: Aboriginal influences and the original state of nature.* Salt Lake City: University of Utah Press.

Kay, C. E., and R. T. Simmons, eds. 2002. *Wilderness and political ecology: Aboriginal influences and the original state of nature.* Salt Lake City: University of Utah Press.

Kershaw, P., S. van der Kaars, P. Moss, and S. Wang. 2002. Quaternary records of vegetation, biomass burning, climate and possible human impact in the Indonesian–Northern Australian region. In P. Kershaw, B. David, N. Tapper, D. Penny, and J. Brown, eds., *Bridging Wallace's line: The environmental and cultural history and dynamics of the SE-Asian-Australian region,* 97–119. Reiskirchen, Germany: Catena Verlag.

King, J. E. 1973. Modern pollen in the Grand Canyon, Arizona. *Geosciences and Man* 7:73–81.

Kirch, P. V. 2000. *On the road of the winds: An archaeological history of the Pacific islands before European contact.* Berkeley: University of California Press.

Kirch, P. V., and T. L. Hunt, eds. 1997. *Historical ecology in the Pacific islands: Prehistoric environmental and landscape change.* New Haven: Yale University Press.

Klein, R. G. 1994. Southern Africa before the Iron Age. In R. S. Corruccini and R. L. Ciochon, eds., *Integrative paths to the past: Paleoanthropological advances in honor of F. Clark Howell*, 471–519. Englewood Cliffs, NJ: Prentice Hall.

———. 1999. *The human career: Human biological and cultural origins.* 2d ed. Chicago: University of Chicago Press.

Koch, P. L., K. A. Hoppe, and S. D. Webb. 1998. The isotopic ecology of late Pleistocene mammals in North America. Part 1: Florida. *Chemical Geology* 152:119–138.

Krech, S., III. 1999. *The ecological Indian: Myth and history.* New York: W. W. Norton.

Kruuk, H. 1972. Surplus killing by carnivores. *Journal of Zoology* 166:233–244.

Kurtén, B. 1988. *Before the Indians.* New York: Columbia University Press.

Kurtén, B., and E. Anderson. 1980. *Pleistocene mammals in North America.* New York: Columbia University Press.

Lange, I. M. 2002. *Ice age mammals of North America: A guide to the big, the hairy, and the bizarre.* Missoula, MT: Mountain Press.

Latorre, C. 1998. Paleontologia de mamíferos del Alero Tres Arroyos I, Tierra del Fuego, XII Region, Chile. Serie C. Nat. Vol. *Anales del Instituto de la Patagonia* 26:77–90.

Laudermilk, J. D., and P. A. Munz. 1934. Plants in the dung of *Nothrotherium* from Gypsum Cave, Nevada. Carnegie Institution of Washington Publication 453:29–37.

Lavallée, D. 2000. *The first South Americans: The peopling of a continent from the earliest evidence to high culture.* Trans. P. G. Bahn. Salt Lake City: University of Utah Press.

Leakey, L. S. B., R. D. Simpson, and T. Clements. 1968. Archaeological excavations in the Calico Mountains, California: Preliminary Report. *Science* 160:1022–1023.

Leakey, R., and R. Lewin. 1995. *The sixth extinction: Patterns of life and the future of humankind.* New York: Doubleday.

Leopold, A. S. 1972. *Wildlife of Mexico: Game birds and mammals.* Berkeley: University of California Press.

Leopold, A. S., S. A. Cain, C. M. Cottam, I. N. Gabrielson, and T. L. Kimball. 1963. Wildlife management in the national parks. *Transactions of the North American Wildlife and Natural Resources Conference* 24:29–44.

Leopold, E. 1967. Late-Cenozoic patterns of plant extinction. In P. S. Martin and H. E. Wright Jr., eds., *Pleistocene extinctions: The search for a cause*, 75–120. New Haven: Yale University Press.

Lister, A., and P. Bahn. 1994. *Mammoths.* New York: Macmillan.

Lister, A. M., and A. V. Sher. 1995. Ice cores and mammoth extinction. *Nature* 378:23–24.

Long, A., R. B. Hendershott, and P. S. Martin. 1983. Radiocarbon dating of fossil eggshell. *Radiocarbon* 24:533–539.

Long, A., and P. S. Martin. 1974. Death of American ground sloths. *Science* 186:638–640.

Long, A., P. S. Martin, and R. M. Hansen. 1974. Extinction of the Shasta ground sloth. *Geological Society of America Bulletin* 85:1843–1848.

Long, A., P. S. Martin, and H. A. Lagiglia. 1998. Ground sloth extinction and human occupation at Gruta del Indio, Argentina. *Radiocarbon* 40:693–700.

Lull, R. S. 1930. The ground sloth, *Nothrotherium*. *American Journal of Science*, 5th series, 20:344–352.

Lyford, M. E., S. T. Jackson, J. L. Betancourt, and S. T. Gray. 2003. Influence of landscape structure and climate variability on a late Holocene plant migration. *Ecological Monographs* 73:567–583.

Lyman, R. L. 1998. *White goats, white lies: The abuse of science in Olympic National Park.* Salt Lake City: University of Utah Press.

Lyons, S. K., F. A. Smith, and J. H. Brown. 2004. Of mice, mastodons, and men: Human-mediated extinctions on four continents. *Evolutionary Ecology Research* 6:339–358.

MacCarter, J. S. 2000. At home in the desert: Oryx thrive too well on White Sands Missile Range. *New Mexico Wildlife* 44 (4):7–9.

MacDonald, G. M. 2003. *Biogeography: Introduction to space, time, and life.* New York: John Wiley & Sons.

MacNeish, R. S., and J. G. Libby, eds. 2003. *Pendejo Cave.* Albuquerque: University of New Mexico Press.

MacPhee, R. D. E., ed. 1999. *Extinctions in near time: Causes, contexts, and consequences.* New York: Kluwer Academic / Plenum Publishers.

MacPhee, R. D. E., C. Flemming, and D. P. Lunde. 1999. "Last occurrence" of the Antillean insectivoran *Nesophontes:* New radiometric dates and their interpretation. *American Museum Novitates* 3261:1–19.

MacPhee, R. D. E., D. C. Ford, and D. A. McFarlane. 1989. Pre-Wisconsinan mammals from Jamaica and models of late Quaternary extinction in the Greater Antilles. *Quaternary Research* 31:94–106.

MacPhee, R. D. E., and P. A. Marx. 1997. The 40,000-year plague: Humans, hyperdisease, and first-contact extinctions. In S. Goodman and B. Patterson, eds., *Natural change and human impact in Madagascar,* 169–217. Washington, DC: Smithsonian Institution Press.

MacPhee, R. D. E., A. N. Tikhonov, D. Mol, C. de Marliave, H. van der Plicht, A. D. Greenwood, C. Flemming, and L. Agenbroad. 2002. Radiocarbon chronologies and extinction dynamics of late Quaternary mammalian megafauna from the Taimyr Peninsula, Russian Federation. *Journal of Archaeological Science* 29:1017–1042.

Madsen, D. B. 2000. A high-elevation Allerod–Younger Dryas megafauna from the west-central Rocky Mountains. *University of Utah Anthropological Papers* 122:100–123.

Manly, W. L. 1987. *Escape from Death Valley.* Reno: University of Nevada Press.

Markgraf, V. 1985. Late Quaternary faunal extinctions in southern Patagonia. *Science* 228:110–112.

Martin, P. S. 1958a. Pleistocene ecology and biogeography of North America. In C. L. Hubbs, ed., *Zoogeography,* 375–420. Publication no. 51. Washington, DC: American Association for the Advancement of Science.

———. 1958b. Taiga-tundra and the full-glacial period in Chester County, Pennsylvania. *American Journal of Science* 256:470–502.

———. 1966. Africa and Quaternary overkill. *Nature* 212:339–342.

———. 1967a. Overkill at Olduvai Gorge. *Nature* 215:212–213.

———. 1967b. Prehistoric overkill. In P. S. Martin and H. E. Wright Jr., eds., *Pleistocene extinctions: The search for a cause*, 75–120. New Haven: Yale University Press.

———. 1969. Wanted: A suitable herbivore. *Natural History* (February):35–39.

———. 1970. Pleistocene niches for alien animals. *Bioscience* 20:218–221.

———. 1973. The discovery of America. *Science* 179:969–974.

———. 1974. Paleolithic players on the American stage. In J. D. Ives and R. Barry, eds., *Arctic and alpine regions*, 669–700. London: Methuen.

———. 1975. Sloth droppings. *Natural History* (August–September): 75–81.

———. 1984. Prehistoric overkill: The global model. In P. S. Martin and R. G. Klein, eds., *Quaternary extinctions: A prehistoric revolution*, 354–403. Tucson: University of Arizona Press.

———. 1986. Refuting late Quaternary extinction models. In D. K. Elliott, ed., *Dynamics of extinction*, 107–129. New York: John Wiley & Sons.

———. 1990. Forty thousand years of extinctions on the "planet of doom." *Palaeogeography, Palaeoclimatology, Palaeoecology* 82:187–201.

———. 1995. Overview: Reflections on prehistoric turbulence. In C. A. Istock and R. S. Hoff, eds., *Storm over a mountain island: Conservation biology and the Mt. Graham affair*, 247–268. Tucson: University of Arizona Press.

———. 1996. Thinking like a canyon: Wild ideas and wild burros. In R. H. Webb, *Grand Canyon, a century of change: Rephotography of the 1889–1890 Stanton expedition*, 82–83. Tucson: University of Arizona Press.

———. 1999. Deep history and a wilder west. In R. Robichaux, ed., *Ecology of Sonoran Desert plants and plant communities*, 255–290. Tucson: University of Arizona Press.

Martin, P. S., and D. Burney. 1999. Bring back the elephants! *Wild Earth* 9:57–64. Reprinted in T. Butler, ed., *Wild earth: Wild ideas for a world out of balance*. Minneapolis: Milkweed Editions.

Martin, P. S., and R. G. Klein, eds. 1984. *Quaternary extinctions: A prehistoric revolution*. Tucson: University of Arizona Press.

Martin, P. S., and P. J. Mehringer Jr. 1965. Pleistocene pollen analysis and biogeography of the Southwest. In H. E. Wright Jr. and D. G. Frey, eds., *The Quaternary of the United States*, 433–451. Princeton: Princeton University Press.

Martin, P. S., B. Sabels, and D. Shutler Jr. 1961. Rampart Cave coprolite and ecology of the Shasta ground sloth. *American Journal of Science* 259:102–127.

Martin, P. S., and F. W. Sharrock. 1964. Pollen analysis of prehistoric human feces: A new approach to ethnobotany. *American Antiquity* 30:168–180.

Martin, P. S., and D. W. Steadman. 1999. Prehistoric extinctions on islands and continents. In R. D. E. MacPhee, ed., *Extinctions in near time: Causes, contexts, and consequences*, 17–55. New York: Kluwer Academic / Plenum Publishers.

Martin, P. S., and C. R. Szuter. 1999. War zones and game sinks in Lewis and Clark's west. *Conservation Biology* 13:36–45.

———. 2002. Game parks before and after Lewis and Clark: Reply to Lyman and Wolverton. *Conservation Biology* 16:244–247.

Martin, P. S., R. S. Thompson, and A. Long. 1985. Shasta ground sloth extinction: A test of the blitzkrieg model. In J. I. Mead and D. J. Meltzer, eds., *En-*

vironments and extinctions: Man in late glacial North America, 5–14. Orono, ME: Center for the Study of Early Man.

Martin, P. S., and H. E. Wright Jr., eds. 1967. *Pleistocene extinctions: The search for a cause.* New Haven: Yale University Press.

Mawby, J. E. 1967. Fossil vertebrates of the Tule Springs site, Nevada. In H. M. Wormington and D. Ellis, eds., *Pleistocene studies in southern Nevada.* Nevada State Museum Anthropological Papers, no. 13. Carson City: Nevada State Museum.

Mayor, A. 2000. *The first fossil hunters: Paleontology in Greek and Roman times.* Princeton: Princeton University Press.

McAuliff, J. R. 1999. The Sonoran Desert: Landscape complexity and ecological diversity. In R. Robichaux, ed., *Ecology of Sonoran Desert plants and plant communities*, 68–115. Tucson: University of Arizona Press.

McDonald, H. G. 1992. Chester's sloths. *Terra* 31:32–33.

———. 2003. Sloth remains from North American caves and associated karst features. In B. W. Schubert, J. I. Mead, and R. W. Graham, eds., *Ice age cave faunas of North America*, 1–16. Bloomington: Indiana University Press.

McFarlane, D. A. 1999a. A comparison of methods for the probabilistic determination of vertebrate extinction chronologies. In R. D. E. MacPhee, ed., *Extinctions in near time: Causes, contexts, and consequences*, 95–103. New York: Kluwer Academic / Plenum Publishers.

———. 1999b. Late Quaternary fossil mammals and last occurrence dates from caves at Barahona, Puerto Rico. *Caribbean Journal of Science* 35:238–248.

McFarlane, D. A., R. D. E. MacPhee, and D. C. Ford. 1998. Body size variability and a Sangamonian extinction model for Amblyrhiza, a West Indian megafaunal rodent. *Quaternary Research* 50:80–89.

McGinnies, W. G. 1981. *Discovering the desert: Legacy of the Carnegie Desert Botanical Laboratory.* Tucson: University of Arizona Press.

McKenna, M. C., and S. K. Bell. 1997. *Classification of mammals above the species level.* New York: Columbia University Press.

McNab, B. K. 2001. Functional adaptations to island life in the West Indies. In C. A. Woods and F. E. Sergile, eds., *Biogeography of the West Indies: Patterns and perspectives*, 55–62. 2d ed. Boca Raton: CRC Press.

Mead, J. I., and L. D. Agenbroad. 1989. Pleistocene dung and extinct herbivores of the Colorado Plateau. *Cranium* 6:29–44.

———. 1992. Isotope dating of Pleistocene dung deposits from the Colorado Plateau, Arizona, and Utah. *Radiocarbon* 34:1–19.

Mead, J. I., L. D. Agenbroad, O. K. Davis, and P. S. Martin. 1986. Dung of *Mammuthus* in the arid Southwest, North America. *Quaternary Research* 25:121–127.

Mead, J. I., L. D. Agenbroad, A. M. Phillips III, and L. T. Middleton. 1987. Extinct mountain goat *(Oreamnos harringtoni)* in southeastern Utah. *Quaternary Research* 27:323–331.

Mead, J. I., L. L. Coats, and B. W. Schubert. 2003. Late Pleistocene faunas from caves in the eastern Grand Canyon, Arizona. In B. W. Schubert, J. I. Mead, and R. W. Graham, eds., *Ice age cave faunas of North America*, 64–87. Bloomington: Indiana University Press.

Mead, J. I., P. S. Martin, R. C. Euler, A. Long, A. J. T. Jull, L. J. Toolin, D. J. Donahue, and T. W. Linick. 1986. Extinction of Harrington's mountain goat. *Proceedings of the National Academy of Sciences* 83:836–839.

Mead, J. I., and D. J. Meltzer. 1984. North American late Quaternary extinctions and the radiocarbon record. In P. S. Martin and R. G. Klein, eds., *Quaternary extinctions: A prehistoric revolution*, 440–450. Tucson: University of Arizona Press.

Mead, J. I., and L. W. Nelson, eds. 1990. *Megafauna and man: Discovery of America's heartland*. Hot Springs: Mammoth Site of South Dakota.

Mead, J. I., M. K. O'Rourke, and T. M. Foppe. 1986. Dung and diet of the extinct Harrington's mountain goat *(Oreamnos harringtoni)*. *Journal of Mammalogy* 67:284–293.

Mead, J. I., and A. M. Phillips III. 1981. The late Pleistocene and Holocene fauna and flora of Vulture Cave, Grand Canyon, Arizona. *Southwestern Naturalist* 26(3):257–288.

Mech, L. D. 1977. Wolf-pack buffer zones as prey reservoirs. *Science* 198:320–321.

Mehringer, P. J., Jr., and C. W. Ferguson. 1969. Pluvial occurrence of bristlecone pine *(Pinus aristata)* in a Mojave Desert mountain range. *Journal of the Arizona Academy of Sciences* 5:284–292.

Mehringer, P. J., and F. F. Foit. 1990. Volcanic ash dating of the Clovis cache at East Wenatchee, Washington. *National Geographic Research* 6:495–503.

Melink, E., and P. S. Martin. 2001. Mortality of cattle on a desert range: Paleobiological implications. *Journal of Arid Environments* 49:671–675.

Meltzer, D. J. 1993. *Search for the first Americans*. Ed. J. A. Sabloff. Washington, DC: Smithsonian Institution Press.

Meretsky, V. J., N. F. Snyder, S. R. Beissinger, D. A. Clendenen, and J. W. Wiley. 2000. Demography of the California Condor: Implications for reestablishment. *Conservation Biology* 14:957–967.

Merrilees, D. 1968. Man the destroyer: Late Quaternary changes in the Australian marsupial fauna. *Journal of the Royal Society of Western Australia* 51:1–24.

Miller, F. L., A. Gunn, and E. Broughton. 1985. Surplus killing as exemplified by wolf predation on newborn caribou. *Canadian Journal of Zoology* 63:295–300.

Miller, G. H., J. W. Magee, B. J. Johnson, M. L. Fogel, N. A. Spooner, M. T. McCulloch, and L. K. Ayliffe. 1999. Pleistocene extinction of *Genyornis newtoni*: Human impact on Australian megafauna. *Science* 283:205–208.

Mithen, S. 1997. Simulating mammoth hunting and extinctions: Implications for North America. In S. van der Leeuw and J. McGlade, eds., *Time, process, and structured transformation in archaeology*, 176–215. London: Routledge.

Molnar, R. E. 2004. *Dragons in the dust: The paleobiology of the giant monitor lizard* Megalania. Bloomington: Indiana University Press.

Monastersky, Richard. 1999. What robbed the Americas of their most charismatic mammals? *Science News* 156 (December 4):360.

Morwood, M. J., P. B. O'Sullivan, F. Aziz, and A. Raya. 1998. Fission-track age of stone tools and fossils on the east Indonesian island of Flores. *Nature* 392:173–176.

Mosimann, J. E., and P. S. Martin. 1975. Simulating overkill by Paleoindians. *American Scientist* 63:304–313.

Moss, C. 1988. *Elephant memories: Thirteen years in the life of an elephant family.* New York: William Morrow.

Mourer-Chauviré, C., R. Bout, S. Ribes, and F. Moutou. 1999. The avifauna of Réunion Island (Mascarene Islands) at the time of the arrival of the first Europeans. In S. L. Olson, ed., *Avian paleontology at the close of the twentieth century: Proceedings of the fourth international meeting of the Society of Avian Paleontology and Evolution, Washington, DC, 4–7 June 1996,* 1–38. Washington, DC: Smithsonian Institution Press.

Mulvaney, D. J., and J. Kamminga. 1999. *The Prehistory of Australia.* Washington, DC: Smithsonian Institution Press.

Mungall, E. C., and W. J. Sheffield. 1994. *Exotics on the range: The Texas example.* College Station: Texas A&M University Press.

Murray, P. 1991. The Pleistocene megafauna of Australia. In P. Vickers-Rich, J. M. Monaghan, R. F. Baird, and T. H. Rich, eds., *Vertebrate paleontology of Australasia,* 1071–1164. Melbourne: Pioneer Design Studio and Monash University Publications.

Murray, P., and P. Vickers-Rich. 2004. *Magnificent mihirungs: The colossal flightless birds of the Australian dreamtime.* Bloomington: University of Indiana Press.

Nelson, L. 1990. *Ice age mammals of the Colorado plateau.* Flagstaff: Quaternary Studies Program, Northern Arizona University.

Nemecek, S. 2000. Who were the first Americans? *Scientific American* 283(3): 80–87.

Norell, M., L. Dingus, and E. Gaffney. 2000. *Discovering dinosaurs: Evolution, extinction, and the lessons of prehistory.* Berkeley: University of California Press.

Nowak, R. M. 1999. *Walker's mammals of the world.* Vols. 1 and 2. Baltimore: Johns Hopkins University Press.

Nunn, P. D. 1994. *Oceanic islands.* Oxford: Blackwell Publishers.

Olson, S. L., and A. Wetmore. 1976. Preliminary diagnoses of two extraordinary new genera of birds from Pleistocene deposits in the Hawaiian islands. *Proceedings of the Biological Society of Washington* 89:247–258.

Olson, S. L., and D. B. Wingate. 2001. A new species of large, flightless rail of the *Rallus longirostris/elegans* complex (Aves: *Rallidae*) from the late Pleistocene of Bermuda. *Proceedings of the Biological Society of Washington* 114:500–516.

Oren, D. C. 1993. Did ground sloths survive to recent times in the Amazonian region? *Goeldiana Zoologia* 19:1–11.

O'Rourke, M. K., and J. I. Mead. 1985. Late Pleistocene and Holocene pollen records from two caves in the Grand Canyon of Arizona, USA. *American Association of Stratigraphic Palynologists* 16:169–186.

Osborn, S. 2002. California Condors: At home in Arizona (conversation of Sophie Osborn with Joshua Brown). *Wild Earth* 12:60–64.

Owen-Smith, R. N. 1988. *Megaherbivores: The influence of very large body size on ecology.* Cambridge: Cambridge University Press.

———. 1989. Megafaunal extinctions: The conservation message from 11,000 years BP. *Conservation Biology* 3:405–412.

———. 1999. The interaction of humans, megaherbivores, and habitats in the late

Pleistocene extinction event. In R. D. E. MacPhee, ed., *Extinctions in near time: Causes, contexts, and consequences*, 57–69. New York: Kluwer Academic / Plenum Publishers.

Peek, J. M. 2000. Mountain goat. In S. Demarais and P. R. Krausman, eds., *Ecology and management of large mammals in North America*, 467–490. Upper Saddle River, NJ: Prentice Hall.

Peterson, K. A. 2003. Human cultural agency in extinction. *Wild Earth* 13:10–14.

Phillips, A. M. 1984. Shasta ground sloth extinction: Fossil packrat midden evidence from the western Grand Canyon. In P. S. Martin and R. G. Klein, eds., *Quaternary extinctions: A prehistoric revolution*, 148–158. Tucson: University of Arizona Press.

Phillips, A., J. Marshall, and G. Monson. 1964. *The birds of Arizona*. Tucson: University of Arizona Press.

Pielou, E. C. 1991. *After the ice age*. Chicago: University of Chicago Press.

Pitulko, V. V., P. A. Nikolsky, E. Yu. Girya, A. E. Basilyan, V. E. Tumskoy, S. A. Koulakov, S. N. Astakhov, E. Yu. Pavlova, and M. A. Anisimov. 2004. The Yana RHS site: Humans in the arctic before the last glacial maximum. *Science* 303:52–59.

Poinar, H., M. Hofreiter, W. G. Spaulding, P. S. Martin, B. A. Stankiewicz, H. Bland, R. P. Evershed, G. Possnert, and S. Pääbo. 1998. Molecular coproscopy: Dung and diet of the extinct ground sloth *Nothrotheriops shastensis*. *Science* 283:402–406.

Poinar, H., M. Kuch, G. McDonald, P. Martin, and S. Pääbo. 2003. Nuclear gene sequences from a late Pleistocene sloth coprolite. *Current Biology* 13:1150–1152.

Porcasi, J. F., T. L. Jones, and L. M. Raab. 2000. Trans-Holocene marine mammal exploitation on San Clemente Island, California: A tragedy of the commons revisited. *Journal of Anthropological Archaeology* 19:200–220.

Porter, S. C. 1989. Some geological implications of average Quaternary conditions. *Quaternary Research* 32:245–261.

Potts, Rick. 1996. *Humanity's descent: The consequences of ecological instability*. New York: William Morrow.

Powell, L. J. 1998. *Night comes to the Cretaceous*. San Diego: Harcourt Brace.

Power, S. 2002. *A problem from hell: America and the age of genocide*. New York: Basic Books.

Pregill, G. K., and B. I. Crother. 1999. Ecological and historical biogeography. In B. I Crother, ed., *Caribbean amphibians and reptiles*, 335–356. San Diego: Academic Press.

Pregill, G. K., and S. L. Olson. 1981. Zoogeography of the West Indies in relation to Pleistocene climatic cycles. *Annual Review of Ecology and Systematics* 12:75–98.

Preston, D. 1995. A reporter at large: The mystery of Sandia Cave. *New Yorker* 71(16):66–83.

Preston, W. L. 2002. Post-Columbian wildlife irruptions in California: Implications for cultural and environmental understanding. In C. E. Kay and R. E. Simmons, eds., *Wilderness and political ecology: Aboriginal influences and the original state of nature*, 111–140. Salt Lake City: University of Utah Press.

Pulliam, H. R. 1988. Sources, sinks, and population regulation. *American Naturalist* 132:652–661.

Putshkov, P. V. 1997. Were the mammoths killed by the warming? (Testing of the climatic versions of Wurm extinctions). *Vestnik Zoologii* supplement 4:3–81.

Pyne, S. J. 1991. *Burning bush: A fire history of Australia.* New York: Henry Holt.

Quammen, D. 1996. *The song of the dodo: Island biogeography in an age of extinctions.* New York: Scribner.

Raup, D. 1991. *Extinction: Bad genes or bad luck?* New York: W. W. Norton.

Rea, A. M. 1998. *Folk mammalogy of the northern Pimas.* Tucson: University of Arizona Press.

Rea, A. M., and L. L. Hargrave. 1984. The bird bones. In R. C. Euler, ed., *The archaeology, geology, and paleobiology of Stanton's Cave, Grand Canyon National Park, Arizona,* 77–91. Grand Canyon Natural History Association, Monograph no. 6. Grand Canyon, AZ: Grand Canyon Natural History Association.

Redman, C. L. 1999. *Human impact on ancient environments.* Tucson: University of Arizona Press.

Reff, D. T. 1991. *Disease, depopulation, and culture change in northwestern New Spain, 1518–1764.* Salt Lake City: University of Utah Press.

Reid, W. H., and G. R. Patrick. 1983. Gemsbok in White Sands National Monument. *Southwestern Naturalist* 28:97–99.

Robbins, E. I., P. S. Martin, and A. Long. 1984. Paleoecology of Stanton's Cave. In R. C. Euler, ed., *The archaeology, geology, and paleobiology of Stanton's Cave, Grand Canyon National Park, Arizona,* 115–130. Grand Canyon Natural History Association, Monograph no. 6. Grand Canyon, AZ: Grand Canyon Natural History Association.

Roberts, R. G., T. F. Flannery, L. K. Ayliffe, H. Yoshida, J. M. Olley, G. J. Prideaux, G. M. Laslett, A. Baynes, M. A. Smith, R. Jones, and B. L. Smith. 2001. New ages for the last Australian megafauna: Continent-wide extinction about 46,000 years ago. *Science* 292:1888–1892.

Robinson, G. S. 2003. Landscape paleoecology and late Quaternary extinctions in the Hudson Valley. Ph.D. diss., Fordham University.

Roosevelt, A. C., J. Douglas, and L. Brown. 2002. The migrations and adaptations of the first Americans: Clovis and pre-Clovis viewed from South America. In N. Jablonski, ed., *The first Americans: The Pleistocene colonization of the New World.* San Francisco: California Academy of Sciences.

Rosenzweig, M. L. 2003. *Win-win ecology: How the earth's species can survive in the midst of human enterprise.* New York: Oxford University Press.

Rostlund, E. 1960. The geographic range of the historic bison in the Southeast. *Annals of the Association of American Geographers* 50:395–407.

Russell, E., and S. D. Stanford. 2000. Late-glacial environmental changes south of the Wisconsin terminal moraine in the eastern United States. *Quaternary Research* 53:105–113.

Russell, S. A. 1996. *When the land was young: Reflections on American archaeology.* Lincoln: University of Nebraska Press.

Salmi, M. 1955. Additional information on the findings in the *Mylodon* Cave at Ultima Esperanza. *Acta Geographica* 14:314–333.

Saunders, J. J. 1977. *Late Pleistocene vertebrates of the western Ozark highland,*

Missouri. Illinois State Museum, Reports of Investigations, no. 33. Springfield: Illinois State Museum.

Sayre, G. 2001. The mammoth: Endangered species or vanishing race? *Journal for Early Modern Cultural Studies* 1:63–87.

Schmidt, G. D., D. W. Duszynski, and P. S. Martin. 1992. Parasites of the extinct Shasta ground sloth, *Nothrotheriops shastensis,* in Rampart Cave, Arizona. *Journal of Parasitology* 78:811–816.

Schubert, B. W. 2003. A late Pleistocene and early Holocene mammalian fauna from Little Beaver Cave, central Ozarks, Missouri. In B. W. Schubert, J. I. Mead, and R. W. Graham, eds., *Ice age cave faunas of North America,* 149–200. Bloomington: Indiana University Press.

Schubert, B. W., R. W. Graham, H. G. McDonald, E. C. Grimm, and T. W. Stafford Jr. 2004. Latest Pleistocene paleoecology of Jefferson's ground sloth (Megalonyx jeffersonii) and elk-moose (Cervalces scottii) in northern Illinois. *Quaternary Research* 61:231–240.

Schubert, B. W., J. I. Mead, and R. W. Graham, eds. 2003. *Ice age cave faunas of North America.* Bloomington: Indiana University Press.

Sellers, W. D., and R. H. Hill. 1974. *Arizona climate 1931–1972.* Rev. 2d ed. Tucson: University of Arizona Press.

Shaw, J. H., and M. Meagher. 2000. Bison. In S. Dermarais and P. R. Krausman, *Ecology and management of large mammals in North America,* 447–466. Upper Saddle River, NJ: Prentice Hall.

Sheridan, T. E. 1995. *Arizona: A history.* Tucson: University of Arizona Press.

Shoshani, J., and P. Tassy, eds. 1996. *The Proboscidea: Evolution and palaeoecology of elephants and their relatives.* Oxford: Oxford University Press.

Simenstad, C. A., J. A. Estes, and K. W. Kenyon. 1978. Aleuts, sea otters, and alternate stable-state communities. *Science* 200:403–411.

Simons, E. L. 1997. Lemurs: Old and new. In S. Goodman and B. Patterson, eds., *Natural change and human impact in Madagascar,* 142–168. Washington, DC: Smithsonian Institution Press.

Simpson, G. G. 1976. *Penguins: Past and present, here and there.* New Haven: Yale University Press.

Sinclair, A. R. E. 1995. Equilibria in plant-herbivore interactions. In A. R. E. Sinclair and P. Arcese, eds., *Serengeti II: Dynamics, management, and conservation of an ecosystem,* 91–113. Chicago: University of Chicago Press.

Smith, M. A., J. R. Prescott, and M. J. Head. 1997. Comparison of ^{12}C and luminescence chronologies at Puritjarra Rock Shelter, Central Australia. *Quaternary Science Reviews* 16:299–320.

Smith, V. L. 1975. The primitive hunter culture, Quaternary extinction and the rise of agriculture. *Journal of Political Economy* 83:717–755.

———. 1999. Economy, ecology and institutions in the emergence of humankind. In L. Barrington, ed., *The other side of the frontier: Economic explorations into Native American history,* 57–85. Boulder: Westview.

Soulé, M. E. 1990. The onslaught of alien species, and other challenges in the coming decade. *Conservation Biology* 4:233–239.

Soulé, M., and R. Noss. 1998. Rewilding and biodiversity: Complementary goals for continental conservation. *Wild Earth* 8:18–28.

Spaulding, W. G. 1990. Vegetational and climatic development of the Mojave Desert: The last glacial maximum to the present. In J. L. Betancourt, T. R. Van Devender, and P. S. Martin, eds., *Packrat middens: The last 40,000 years of biotic change*, 166–199. Tucson: University of Arizona Press.

Spaulding, W. G., and P. S. Martin. 1979. Ground sloth dung of the Guadalupe Mountains. In H. H. Genoways and R. J. Baker, eds., *Biological investigations in the Guadalupe Mountains National Park, Texas*, 259–269. U.S. National Park Service Proceedings and Transactions Series, no. 4. Washington, DC: Department of the Interior, National Park Service.

Spaulding, W. G., and K. L. Petersen. 1980. Late Pleistocene and early Holocene paleoecology of Cowboy Cave. In J. D. Jennings, ed., *Cowboy Cave*, 163–177. Anthropological Papers, University of Utah, no. 104. Salt Lake City: University of Utah Press.

Speth, J. 1983. *Bison kills and bone counts: Decision making by ancient hunters*. Chicago: University of Chicago Press.

Stacey, M. H. 1929. *Uncle Sam's camels: The journal of May Humphreys Stacey*. Ed. L. B. Lesley. Cambridge, MA: Harvard University Press.

Stafford, T. W., P. E. Hare, L. Currie, A. C. T. Jull, and D. J. Donahue. 1991. Accelerator radiocarbon dating at the molecular level. *Journal of Archaeological Science* 18:35–72.

Steadman, D. W. 1995. Prehistoric extinctions of Pacific island birds: Biodiversity meets zooarchaeology. *Science* 267:1123–1131.

———. 1997. Extinctions of Polynesian birds: Reciprocal impacts of birds and people. In P. V. Kirch and T. L. Hunt, eds., *Historical ecology in the Pacific islands: Prehistoric environmental and landscape change*, 51–79. New Haven: Yale University Press.

———. n.d. *Biogeography and extinction of tropical Pacific birds*. Chicago: University of Chicago Press. Forthcoming.

Steadman, D. W., and P. S. Martin. 1984. Extinction of birds in the late Quaternary of North America. In P. S. Martin and R. G. Klein, eds., *Quaternary extinctions: A prehistoric revolution*, 466–477. Tucson: University of Arizona Press.

———. 2003. The late Quaternary extinction and future resurrection of birds on Pacific islands. *Earth-Science Reviews* 61:133–147.

Steadman, D. W., G. K. Pregill, and D. V. Burley. 2002. Rapid prehistoric extinction of iguanas and birds in Polynesia. *Proceedings of the National Academy of Sciences* 99:3673–3677.

Steadman, D. W., and S. Zousmer. 1988. *Galapagos: Discovery on Darwin's islands*. Washington, DC: Smithsonian Institution Press.

Sterling, B. 2002. *Tomorrow now: Envisioning the next fifty years*. New York: Random House.

Stiner, M. C., N. D. Munroe, and T. A. Surovell. 2000. The tortoise and the hare: Small game use, the broad-spectrum revolution and Paleolithic demography. *Current Anthropology* 41:39–79.

Stone, R. 2001. *Mammoth: The resurrection of an ice age giant*. Cambridge: Perseus Publishing.

Stuart, A. J. 1991. Mammalian extinctions in the late Pleistocene of northern Eurasia and North America. *Biological Reviews* 66:453–562.

————. 1999. Late Pleistocene megamammal extinctions: A European perspective. In R. D. E. MacPhee, ed., *Extinctions in near time: Causes, contexts, and consequences*, 257–269. New York: Kluwer Academic / Plenum Publishers.

Stuart, A. J., L. D. Sulerzhitsky, L. A. Orlova, Y. V. Kuzmin, and A. M. Lister. 2002. The latest woolly mammoths (*Mammuthus primigenius* Blumenbach) in Europe and Asia: A review of the current evidence. *Quarternary Science Reviews* 21:1559–1569.

Sukumar, R. 1994. *Elephant days and nights*. New York: Oxford University Press.

Sukumar, R., and C. Santiapillai. 1996. *Elephas maximus*: Status and distribution. In J. Shoshani and P. Tassy, eds., *The Proboscidea: Evolution and palaeoecology of elephants and their relatives*, 327–331. Oxford: Oxford University Press.

Surovell, T. A. 2000. Early Paleoindian women, children, mobility, and fertility. *American Antiquity* 65:493–509.

Surovell, T. A., N. M. Waguespack, and P. J. Brantingham. 2005. Global archaeological evidence for proboscidean overkill. *Proceedings of the National Academy of Science* 102:6231–6236.

Sutcliffe, A. 1985. *On the track of ice age mammals*. Cambridge, MA: Harvard University Press.

Szuter, C. R. 1991. *Hunting by prehistoric horticulturists in the American Southwest*. New York: Garland Publishing.

Tankersley, K. B. 2002. *In search of ice age Americans*. Salt Lake City: Gibbs Smith.

Taylor, K. C., G. W. Lamorey, and G. A. Doyle. 1993. The "flickering switch" of late Quaternary climate change. *Nature* 361:432–436.

Taylor, R. E. 1987. *Radiocarbon dating: An archaeological perspective*. Orlando: Academic Press.

Taylor, R. E., C. V. Haynes, and M. Stuiver. 1996. Calibration of the late Pleistocene radiocarbon time scale: Clovis and Folsom age estimates. *Antiquity* 70:515–525.

Terrell, John. 1986. *Prehistory of the Pacific islands: A study of variation in language, customs, and human biology*. New York: Cambridge University Press.

Thomas, D. H. 2000. *Skull wars: Kennewick man, archaeology, and the battle for Native American identity*. New York: Basic Books.

Thompson, R. S. 1990. Late Quaternary vegetation and climate in the Great Basin. In J. L. Betancourt, T. R. Van Devender, and P. S. Martin, eds., *Packrat middens: The last 40,000 years of biotic change*, 200–239. Tucson: University of Arizona Press.

Thompson, R. S., T. R. Van Devender, P. S. Martin, T. Foppe, and A. Long. 1980. Shasta ground sloth (*Nothrotheriops shastense* Hoffstetter) at Shelter Cave, New Mexico: Environment, diet, and extinction. *Quaternary Research* 14(3): 360–376.

Tonni, E. P., and R. C. Pascuali. 1998. *Mamíferos fósiles: Cuando en la pampa Vivian los Gigantes*. La Plata, Argentina: Museo Nacional de Historia Natural.

Truett, J. 1996. Bison and elk in the American Southwest: In search of the pristine. *Environmental Management* 20:195–206.

Turner, R. M., R. H. Webb, J. E. Bowers, and J. R. Hastings. 2003. *The chang-

ing mile revisited: An ecological study of vegetation change with time in the lower mile of an arid and semiarid region. Tucson: University of Arizona Press.

Turney, C. S. M., A. P. Kershaw, P. Moss, M. I. Bird, L. K. Fifield, R. G. Cresswell, G. M. Santos, M. L. De Tada, P. A. Hausladen, and Y. Zhou. 2001. Redating the onset of burning at Lynch's Crater (North Queensland): Implications for human settlement in Australia. *Journal of Quaternary Science* 16:767–771.

Valladas, H., J. Clottes, J. M. Geneste, M. A. Garcia, M. Arnold, H. Cachier, and N. Tisnérat-Laborde. 2001. Evolution of prehistoric cave art. *Nature* 413:479.

Van Devender, T. R., P. S. Martin, A. M. Phillips III, and W. G. Spaulding. 1977. Late Quaternary biotic communities from the Guadalupe Mountains, Culberson County, Texas. In R. H. Wauer and D. H. Riskind, eds., *Transactions of the Symposium on the biological resources of the Chihuahuan Desert Region, United States and Mexico*, 107–113. U.S. National Park Service Proceedings and Transactions Series, no. 3. Washington, DC: Department of the Interior, National Park Service.

Van Devender, T. R., and J. I. Mead. 1976. Late Quaternary and modern plant communities of Shinumo Creek and Peach Springs Wash, Lower Grand Canyon, Arizona. *Journal of the Arizona Academy of Science* 11:16–22.

Van de Water, P. K., S. W. Leavitt, and J. L. Betancourt. 1994. Trends in stomatal density and $^{13}C/^{12}C$ ratios of *Pinus flexilis* needles during last glacial-interglacial cycle. *Science* 264:239–243.

Vartanyan, S. L., V. E. Garutt, and A. V. Sher. 1992. Holocene dwarf mammoths from Wrangel Island in the Siberian Arctic. *Nature* 362:337–340.

Vaughn, T. A. 1990. Ecology of living packrats. In J. L. Betancourt, T. R. Van Devender, and P. S. Martin, eds., *Packrat middens: The last 40,000 years of biotic change*, 14–27. Tucson: University of Arizona Press.

Vera, F. W. M. 2000. *Grazing ecology and forest history.* New York: CAB International.

Vilà, C., J. A. Leonard, A. Götherström, S. Marklund, K. Sandberg, K. Lidén, R. K. Wayne, and H. Ellegren. 2001. Widespread origins of domestic horse lineages. *Science* 291:474–477.

Vrba, E. S. 1995. On the connections between paleoclimate and evolution. In E. S. Vrba, G. H. Denton, T. C. Partridge, and L. H. Burckle, eds., *Paleoclimate and evolution with emphasis on human origins*, 24–48. New Haven: Yale University Press.

———. 1999. Habitat theory in relation to the evolution in African Neogene biota and hominids. In T. G. Bromage and F. Schrenk, eds., *African biogeography, climate change and human evolution*, 19–34. New York: Oxford University Press.

Wagner, F. H., and C. E. Kay. 1993. "Natural" or "healthy" ecosystems: Are U.S. national parks providing them? In M. J. McDonnell and S. T. Pickett, eds., *Humans as components of ecosystems*, 257–270. New York: Springer-Verlag.

Waguespack, N. M., and T. A. Surovell. 2003. Clovis strategies, or how to make out on plentiful resources. *American Antiquity* 68:333–352.

Wallace, D. R. 2004. Beasts of Eden: Walking whales, dawn horses, and other enigmas of mammal evolution. Berkeley: University of California Press.

Ward, P. 1994. *The end of evolution: On mass extinctions and the preservation of biodiversity.* New York: Bantam Books.

———. 1997. *The call of distant mammoths: Why the ice age mammals disappeared.* New York: Copernicus Books.

———. 2000. *Rivers in time: The search for clues to earth's mass extinctions.* New York: Columbia University Press.

Watts, W. A. 1980. Regional variation in the response of vegetation to late-glacial climatic events in Europe. In J. J. Lowe, J. M. Gray, and J. E. Robinson, eds., *Studies in the late-glacial of northwest Europe*, 1–22. Oxford: Pergamon.

Webb, R. H. 1996. *Grand Canyon, a century of change: Rephotography of the 1889–1890 Stanton expedition.* Tucson: University of Arizona Press.

Webb, S. D. 1984. Ten million years of mammalian extinctions in North America. In P. S. Martin and R. G. Klein, eds., *Quaternary extinctions: A prehistoric revolution*, 189–210. Tucson: University of Arizona Press.

Webb, S. D., R. W. Graham, A. D. Barnowski, C. J. Bell, R. Franz, E. A. Hadley, E. L. Lundelius Jr., H. G. McDonald, R. A. Martin, H. A. Semken Jr., and D. W Steadman. 2004. Vertebrate paleontology. In A. R. Gillespie, S. Porter, and B. F. Atwater, eds., *The Quaternary period in the United States*, 519–538. Amsterdam: Elsevier.

Webb, S. D., and A. Rancy. 1996. Late Cenozoic evolution of the Neotropical mammal fauna. In J. B. C. Jackson, A. F. Budd, and A. G. Coates, eds., *Evolution and environment in tropical America.* Chicago: University of Chicago Press.

Wells, P. V., and C. D. Jorgensen. 1964. Pleistocene wood rat middens and climatic change in Mohave Desert: A record of juniper woodlands. *Science* 143:1171–1174.

Weng, C., and S. T. Jackson. 1999. Late glacial and Holocene vegetation history and paleoclimate of the Kaibab Plateau, Arizona. *Palaeogeography, Palaeoclimatology, Palaeoecology* 153:179–201.

Western, D. 1997. *In the dust of Kilimanjaro.* Washington, DC: Island Press.

White, J. L., and R. D. E. MacPhee. 2001. The sloths of the West Indies: A systematic and phylogenetic review. In C. A. Woods and F. E. Sergile, eds., *Biogeography of the West Indies: Patterns and perspectives*, 201–236. 2d ed. Boca Raton: CRC Press.

Whittington, S. L., and B. Dyke. 1984. Simulating overkill: Experiments with the Mosimann and Martin model. In P. S. Martin and R. G. Klein, eds., *Quaternary extinctions: A prehistoric revolution*, 451–465. Tucson: University of Arizona Press.

Wiles, G. J., J. Bart, R. E. Beck Jr., and C. F. Aguon. 2003. Impacts of the brown tree snake: Patterns of decline and species persistence in Guam's avifauna. *Conservation Biology* 17:1350–1360.

Williams, G. W. 2002. Aboriginal use of fire: Are there any natural plant communities? In C. E. Kay and R. E. Simmons, eds., *Wilderness and political ecology: Aboriginal influences and the original state of nature*, 179–214. Salt Lake City: University of Utah Press.

Wilson, E. O. 1992. *The diversity of life.* Cambridge, MA: Belknap Press.

———. 2002. *The future of life.* New York: Alfred A. Knopf.

Wilson, R. W. 1942. Preliminary study of the fauna of Rampart Cave, Arizona.

In Carnegie Institution of Washington, *Studies of Cenozoic vertebrates of western North America and of fossil primates,* 169–185. Carnegie Institution of Washington Publication no. 530. Washington, DC: Carnegie Institution.

Winterhalder, B., and F. Lu. 1997. A forager-resource population ecology model and implications for indigenous conservation. *Conservation Biology* 11:1354–1364.

Woodford, J. 2000. *The Wollemi pine: The incredible discovery of a living fossil from the age of the dinosaurs.* Melbourne: Text Publishing.

Wormington, H. M. 1957. *Ancient man in North America.* 4th ed. Denver: Museum of Natural History.

Wormington, H. M., and D. Ellis, eds. 1967. *Pleistocene studies in southern Nevada.* Nevada State Museum Anthropological Papers, no. 13. Carson City: Nevada State Museum.

Worthy, T. H. 1999. The role of climate change versus human impacts: Avian extinction on South Island, New Zealand. *Smithsonian Contributions to Paleobiology* 89:111–123.

Worthy, T. H., and R. N. Holdaway. 2002. *The lost world of the moa: Prehistoric life of New Zealand.* Bloomington: Indiana University Press.

Wroe, S., J. Field, R. Fullagar, and L. S. Jermiin. 2004. Megafaunal extinction in the late Quaternary and the global overkill hypothesis. *Alcheringia* 28:291–331.

Wroe, S., T. Myers, F. Seebacher, B. Kear, A. Gillespie, M. Crowther, and S. Salisbury. 2003. An alternative method for predicting body mass: The case of the Quaternary marsupial lion. *Paleobiology* 29:403–411.

Wuerthner, G., and M. Matteson. 2002. *Welfare ranching: The subsidized destruction of the American West.* Washington, DC: Island Press.

Zimov, S. A. 2005. Pleistocene park: return of the mammoth's ecosystem. *Science* 308:796–798.

Zimov, S. A., V. I. Chuprynin, A. P. Oreshko, F. S. Chapin, J. F. Reynolds, M. C. Chapin. 1995. Steppe-tundra transition: A herbivore-driven biome shift at the end of the Quaternary. *American Naturalist* 146:765–793.

Index

Indexer:	Andrew Christenson
Compositor:	Integrated Composition Systems
Text:	Sabon
Display:	Franklin Gothic
Printer and Binder:	Thomson-Shore, Inc.